Adela

Manufacturing system simulation & human robot interaction in VR

Adelaide Marzano

Manufacturing system simulation & human robot interaction in VR

Methodologies, software tools, case study in transportation design

VDM Verlag Dr. Müller

Impressum/Imprint (nur für Deutschland/ only for Germany)

Bibliografische Information der Deutschen Nationalbibliothek: Die Deutsche Nationalbibliothek verzeichnet diese Publikation in der Deutschen Nationalbibliografie; detaillierte bibliografische Daten sind im Internet über http://dnb.d-nb.de abrufbar.

Alle in diesem Buch genannten Marken und Produktnamen unterliegen warenzeichen-, marken- oder patentrechtlichem Schutz bzw. sind Warenzeichen oder eingetragene Warenzeichen der jeweiligen Inhaber. Die Wiedergabe von Marken, Produktnamen, Gebrauchsnamen, Handelsnamen, Warenbezeichnungen u.s.w. in diesem Werk berechtigt auch ohne besondere Kennzeichnung nicht zu der Annahme, dass solche Namen im Sinne der Warenzeichen- und Markenschutzgesetzgebung als frei zu betrachten wären und daher von jedermann benutzt werden dürften.

Coverbild: www.purestockx.com

Verlag: VDM Verlag Dr. Müller Aktiengesellschaft & Co. KG
Dudweiler Landstr. 99, 66123 Saarbrücken, Deutschland
Telefon +49 681 9100-698, Telefax +49 681 9100-988, Email: info@vdm-verlag.de
Zugl.: Naples, University of Naples Federico II, Diss., 2008

Herstellung in Deutschland:
Schaltungsdienst Lange o.H.G., Berlin
Books on Demand GmbH, Norderstedt
Reha GmbH, Saarbrücken
Amazon Distribution GmbH, Leipzig
ISBN: 978-3-639-19757-0

Imprint (only for USA, GB)

Bibliographic information published by the Deutsche Nationalbibliothek: The Deutsche Nationalbibliothek lists this publication in the Deutsche Nationalbibliografie; detailed bibliographic data are available in the Internet at http://dnb.d-nb.de .

Any brand names and product names mentioned in this book are subject to trademark, brand or patent protection and are trademarks or registered trademarks of their respective holders. The use of brand names, product names, common names, trade names, product descriptions etc. even without a particular marking in this works is in no way to be construed to mean that such names may be regarded as unrestricted in respect of trademark and brand protection legislation and could thus be used by anyone.

Cover image: www.purestockx.com

Publisher:
VDM Verlag Dr. Müller Aktiengesellschaft & Co. KG
Dudweiler Landstr. 99, 66123 Saarbrücken, Germany
Phone +49 681 9100-698, Fax +49 681 9100-988, Email: info@vdm-publishing.com

Printed in the U.S.A.
Printed in the U.K. by (see last page)
ISBN: 978-3-639-19757-0

ADELAIDE MARZANO

MANUFACTURING SYSTEM SIMULATION AND HUMAN ROBOT INTERACTION IN VIRTUAL REALITY

"Sometimes dreams come true!
This important task falls upon you, with
your devotion, persistence and passion".

CONTENTS

CHAPTER 4

MANUAL WORK CELL SIMULATION IN A VIRTUAL ENVIRONMENT

CHAPTER 5

**AUTOMATIC ASSEMBLY LINE SIMULATION IN A VIRTUAL
ENVIRONMENT**

CHAPTER 6

HUMAN ROBOT INTERACTION IN VIRTUAL REALITY

INTRODUCTION

Virtual Reality techniques are relatively new, having experienced significant development only during the last few years, in accordance with the progress achieved by computer science and hardware and software technologies. Therefore, there has not yet been a great diffusion and application of VR techniques in the industrial field, in spite of the constant reduction in the costs. This is particularly true in Italy, the subject of this discussion. The study of such advanced design systems has led to the realization of an immersive environment in which new procedures for the evaluation of product prototypes, ergonomics and manufacturing operations have been simulated. The positive and enthusiastic response received from the industrial world has confirmed that these methodologies can be extremely useful in the design phase, influencing the development time and the quality of industrial products.

The application of the environment realized to robotics, ergonomics, plant simulations and maintainability verifications has allowed us to highlight the advantages offered by a design methodology: the possibility of working on the industrial product in the first phase of conception; of placing the designer in front of the virtual reproduction of the product in a realistic way; and of interacting with the same concept. All this allows the modification and improvement of the product characteristics in real time with a remarkable saving of time and costs. Moreover, during the application of VR to industrial case studies, the designers could take advantage of the developed methodology in the design phase, in particular in the planning of new service systems, having the possibility to visualize and to interact with models in real dimensions.

The research area "Virtual Manufacturing" (hereafter often abbreviated as VM) can be defined as an integrated manufacturing environment which can enhance one or more levels of decision and control in the manufacturing process. Several domains can be addressed: Product and Process Design, Process and Production Planning, Machine Tools and Robot and Manufacturing Systems. As automation technologies such as CAD/CAM have substantially shortened the time required to design products, VM is having a similar effect on the manufacturing phase thanks to the modelling, simulation and optimisation of the product and the processes involved in its fabrication.

Manufacturing is an indispensable part of the economy and is the central activity that encompasses product, process, resources and plant. Nowadays products are more and more complex, processes are highly sophisticated and use micro-technology and mechatronics and market demand is evolving and expanding rapidly, so that we need a flexible and lean production. Moreover manufacturing enterprises may be widely distributed geographically and linked conceptually in terms of dependence and material, information and knowledge flow. In this complex and developing environment, industrialists must be informed about their processes before their application in order to "get it right first time". To achieve this goal, the use of a VM environment provides a computer-based environment to simulate individual manufacturing processes and the total manufacturing enterprise. VM systems enable early optimization of cost, quality and time drivers, achieve integrated product, process and resource design and finally allow an early consideration of productivity and affordability.

The aim of this research activity is to present an updated vision of VM through different aspects. This study will take into account the market penetration of several tools with respect to their state of development and the differences in terms of effort and level of detail between industrial tools and academic research. We will describe the trends and results achieved in the automotive, aerospace and railway fields, in terms of the Digital Product Creation Process to design the product and the manufacturing process.

Chapter 1 defines the objectives and the scope of the VM domain. The expected technological benefits of VM will also be presented. Chapter 2 briefly introduces Virtual Reality and describes the VR laboratory, named "VRTest", realized for the Competence Regional Centre for the qualification of transportation systems set up by Campania Regional Authority. Chapter 3 describes a complete Virtual Reality environment realized to simulate manufacturing systems. An original VR architecture has been conceived in order to create a unique environment, the features of which are able to satisfy the requirements of Virtual Manufacturing tasks. In Chapter 4 two case studies are described concerning manual work cell simulation. In particular, the first deals with a developed methodology for studying in a virtual environment the ergonomics of a work cell in an automotive manufacturing system; the second deals with an ergonomic evaluation through a direct manual interaction approach in VR. Chapter 5 deals with automatic assembly line simulations in a virtual environment. Two case studies are presented relating to applications that concern the use of simulation methodologies within the manufacturing process, one in the railway, the other in the aerospace field. Chapter 6 describes a framework which helps designers visualize and verify the results of robotic work cell simulation in a virtual environment. The same framework allows us to verify interface usability for robot control. The development of a simulation environment, where three different manipulators can be mounted on a commercially available wheelchair, is considered. Experimental results are discussed in a significant case study, based on users' feedback.

The most important advantages deriving from the application of Virtual Reality techniques in product design are often summarized by means of the formula "developing time and costs saving". This statement highlights the effective profitable consequences for a company of the introduction and the systematic application of VR in the development cycle of a new product. The possibility of designing an alternative product solution just involving the creativity of a designer and the costs of his working time and the technologies employed, saying nothing about design in "free time", represents a revolution in terms of development, variety and quality of the product.

CHAPTER 1

DIGITAL MANUFACTURING vs VIRTUAL MANUFACTURING

1.1 WHAT IS DIGITAL MANUFACTURING?

Digital Manufacturing (hereafter often abbreviated DM) can be defined as a proven software-based solution that supports effective collaborative manufacturing process planning between engineering disciplines, such as design and manufacturing. This requires access to the full digital product definition, including tooling and manufacturing process data. Integrated tool suites working off this product definition are used to support visualization, simulation, and other analyses necessary to optimize the product and manufacturing process design. Requirements from across different engineering disciplines are also supported through this process.

As shown in Figure 1.1, the six major DM functions typically include:

• Data synchronization from design through manufacturing in an enterprise information management environment, including linkage and data integration among CAD, CAM, tool design, ERP, MES, and other software applications.

• A systematic, structured, visual, and analytical approach to part and assembly computer-aided process planning to obtain an optimal process solution. Establishing and cataloguing manufacturing constraints, costs, throughputs, and best practices is also performed.

• Detailed line, cell, station, and task design for part manufacturing and assembly process management, including plant design and creation of mechanical assembly-line layouts.

• Discrete event simulation of manufacturing operations and material flows to visualize, validate, and optimize processes, including production line balancing, measurement and verification of line performance. Simulation and assessment of worker movement, ergonomics, safety and performance is provided to assure compliance with government and industry standards.

• Maintaining and managing information on manufacturing resources, including software to support commonization and re-use of parts, assemblies, equipment, and processes. The software also provides manufacturing documentation, shop floor instruction, improved visualization, effective communication, and collaboration among workers.

• Programming of robots, welding, painting, coordinate measuring machines, and other factory equipment, as well as creation, testing, optimizing, and managing printed circuit boards and product assemblies. Quality planning, product inspection, control of dimensional variation, and continual assessment of production quality is also provided.

12

Figure 1.1. Digital Manufacturing functions.

1.2 THE ROLE OF DIGITAL MANUFACTURING

DM software-based solutions are utilized by manufacturing engineers to determine how to build products, synchronize engineering and manufacturing operations, and unify the production environment.

An optimized production environment is established by more automated and effective process planning, plant design, and workflow simulation; the cost of inventory, direct labor, manufacturing engineering, plant and equipment is reduced. A common digital model is employed in Digital Manufacturing to bring together information on products, processes, plants, tools, and resources to enhance operational efficiency, establish best practices, and provide a consistent manufacturing solution. Utilization of a digital model provides a basis for collaboration, synchronization of processes, and highly efficient workflows.

DM is an integral component of a PLM (Product Lifecycle Management) solution (Figure 1.2). Maximum integration of product design, manufacturing engineering, and production operations is achieved and effective use of information is established up and down a supply chain. CPDM (Collaborative Product Data Management) capabilities for collaboration, change management, document management, workflow management, process management, resource management, and version and product variation management can be implemented and integrated with Digital Manufacturing to create a highly effective enterprise-wide product lifecycle solution. Digital Manufacturing's domain of support within the overall digital product lifecycle is illustrated in Figure 1.2.

1.3   DIGITAL MANUFACTURING IN PLM

In today's hyper-competitive environment, manufacturing organizations continually search for, develop, and fund projects that will reduce manufacturing costs and increase productivity.

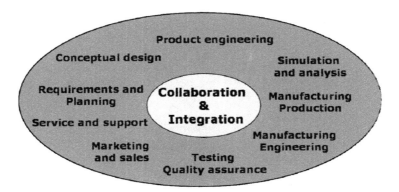

Figure 1.2. Digital Manufacturing functions.

However, these traditional approaches are quickly reaching diminishing returns in their ability to reduce costs and improve productivity. A new approach is needed. DM is that new approach that substitutes moving virtual bits for moving physical atoms. DM focuses on the processes that realize the virtual product. While DM can provide benefits when used as a stand-alone initiative, it is most effectively used when it is incorporated into a broad enterprise Product Lifecycle Management (PLM) initiative. Extending PLM processes and practices, including Digital Manufacturing, to involve suppliers is particularly compelling and should result in substantially reduced costs, more innovation, and much higher benefits. DM offerings combined with data management and CAD tools provide a comprehensive suite of solutions that integrate the definitions of products, processes, practices, plants, tools, and resources into a consistent manufacturing solution. Integrating the product design process with product data provides a full solution that supports processes throughout this specific focus of the lifecycle, from concept-to-production.

DM is a departure from traditional methods of manufacturing planning, control, and implementation. As such it requires different people skills, different business processes and practices, and the right DM technology. For organizations in competitive industry sectors embracing DM will not simply be required in order to be competitive. DM will be required for survival. The key elements for a successful DM initiative will be:

• Understanding what DM is;

• Developing a DM strategy to support specific projects and research activity;

• Quantifying the financial impact of the specific projects and research activity.

1.4 DIGITAL MANUFACTURING DOMAIN

Technologies and processes used by companies to design their products over the past thirty years have proven to be so successful and have advanced to the point that the design information can now be used to streamline the production of the actual product. Thus, a variety of technologies and methods that support manufacturing process design have been developed and implemented over the past several years. Until recently, these efforts typically have been point solutions directed primarily towards manufacturing engineering organizations and usually not fully integrated into broader product lifecycle management (PLM) environments that support product design activities. New DM solutions work within enterprise-wide PLM frameworks to support the development of improved product production processes providing essential support for lean manufacturing initiatives. The use of PLM to support DM provides the potential for substantial benefits. Today's DM solutions enhance and expand upon the capabilities of the long established manufacturing engineering technologies used in the past, including CAM, machine tool simulation, and robot simulation. In contrast to these tools that support specific processes, DM supports collaboration across the entire product design and manufacturing process allowing industrial companies to transform how they develop their manufacturing processes. DM within a PLM context enables collaboration among manufacturing engineers and design engineers on a much greater scale than has previously been achieved, thus fostering innovation in products and in product-related process definition and utilizing the value of the organization's rich product development data. It is imperative that DM solutions exist within a PLM environment that can manage the processes, data, and resources across the production planning activities and among partners, suppliers, and customers. While DM impacts the overall product lifecycle by fully integrating the definitions of products, processes, and resources into a comprehensive and consistent manufacturing solution, it is primarily focused on supporting the portion of the lifecycle that is centred around manufacturing planning and manufacturing engineering activities. It also simplifies the boundaries between PLM and enterprise resource planning (ERP), two primary areas of IT investment for industrial companies, facilitating the development of integrations between these two important enterprise solutions. Extending PLM processes, including DM, to involve suppliers is particularly compelling and should result in substantially improved operations, more innovation, and much higher benefits. DM offerings combined with data management and CAD tools provide a comprehensive suite of solutions that integrate the definitions of products, processes, plants, tools, and resources into a consistent manufacturing solution. Integrating the manufacturing design process with product data provides a full solution that supports processes throughout the lifecycle, from concept-to-production. By involving the extended enterprise including product design, process design, quality, suppliers, and management, companies have been able to achieve substantial benefits throughout the enterprise. The culture of the organization may be as important as the technology that it uses. To achieve the largest benefits, companies have to cross over the traditional boundaries in their organizations. The breadth of Digital Manufacturing solutions has continued to evolve and mature. Today, full manufacturing facility and process definition and simulation, encompassing tooling, assembly lines, work centre, ergonomics, and resource planning, are an integral part of the manufacturing engineering environment. Simulations of all facets of the production process can be used to optimize manufacturing processes to gain a more complete understanding of the production of products, including machine operations and human interaction during assembly. Feedback from actual production operations can be incorporated and used to modify process and resource definitions to take maximum advantage of real experiences and better utilize facilities and resources. The virtual product definition can now be seen as a full simulation of the definition and creation of the product, from raw materials through final assembly, made possible by including processes, resources, product definition data flow, interactions with partners, and process simulations. In addition, while DM is certainly beneficial by decreasing the costs of wasted time, energy, and material, it can also have a major impact on the revenue side of the equation. DM enables increasing complexity, decreasing cycle times, globalization, and regulatory compliance. However, DM will also enable companies to continue to increase their capacity by improving the focus on innovation, collaboration, and quality. Planning for DM and PLM is very important. Work processes need to be refined and proper infrastructure needs to be in place to gain the highest level of benefits from these technologies and methods.

1.5 DIGITAL MANUFACTURING SOFTWARE-BASED OBJECTIVES

By implementing DM software-based solutions, producers around the world are placing themselves in a stronger position to meet the forces encountered in an increasingly demanding and challenging global manufacturing environment. The relentless pressures to continuously improve productivity, lower costs, compress delivery times, and enhance quality of products must be met, while at the same time, internal business objectives must be achieved. Enlightened firms are incorporating more efficient processes, DM software-based solutions, and other technologies to create an efficient and lean production operation and prosper on the world stage. DM solutions are based on an integrated set of software capabilities that utilize a digital product model and work with product definition data to support part and assembly planning, process

16

design, visualization, simulation and other analyses to digitally plan, validate and optimize a manufacturing process. Through implementation, manufacturers are addressing their business needs by:

• Continuously improving the efficiency of manufacturing processes to ensure production flexibility, high performance and superior quality

• Commonizing and re-using parts, assemblies, equipment, and processes

• Assurance that compliance with appropriate industry and government standards is being met

• Managing and synchronizing product and process information from idea conception to the end of product life

The benefits of Digital Manufacturing have been clearly demonstrated in many successful implementations around the world and include:

• Shortened development cycles

• Reduced production costs and improved quality

• Support of lean manufacturing and agility

• Enabled DFx initiatives

• Support of product knowledge dissemination

1.6 PRESSURES ON MANUFACTURERS

Manufacturers are faced with increasingly intense internal and external pressures from shareholders and the market. Shareholders demand steadily-growing revenues and profitability on a quarter-by-quarter basis while the market expects product innovation and aesthetic appeal; greater product functionality, performance, and usability; and a longer, useful life for products. Producers are caught in this confluence of forces that places an onerous squeeze on executives, managers, planners, and workers to lower costs, shorten lead times, and improve quality in order to effectively compete in a worldwide economy.

Globalization has increased the stress placed on producers. Outsourcing, insourcing, and offshoring have become ubiquitous, and manufacturers in all areas of the world now compete aggressively for end user buyers or to be an integral component in a product supply chain. Those competing in global manufacturing are continually faced with a series of challenges such as:

• Limited resources of people, time and funding

• Production risk and a lack of confidence in being able to meet program objectives in a timely manner

• An excessive amount of time spent locating relevant information and ineffective oral, written and electronic communications within an operation, an enterprise and a supply chain

• Engineers and production personnel having difficulty visualizing and optimizing a future production process

• Discovery of design or manufacturing problems late in the cycle when the cost and time impact is the greatest and spending too much time and money in re-design and re-working a production process to achieve program and business objectives.

Global supply chains are increasingly critical to a firm's well-being. They typically include multiple firms from multiple geographies throughout the world. Manufacturing is a worldwide race favoring those firms that are cost-efficient, agile, lean, structured, process-oriented, technology oriented, and have created a culture that incessantly strives for excellence. Competing in a global workplace demands continuous improvements in processes, operations, people, and technology. Manufacturing productivity must be ever increased and producers must constantly look for ways to meet the faster, better, cheaper mantra of today's economy. To meet these pressures and remain competitive, leading manufacturers are going digital. By going digital, producers:

• Can synchronize data and minimize data re-entry and translation through integrated application solutions

• Share and collaborate on intellectual property within a supply chain

• Visualize and digitally simulate operations

• Optimize cells, assembly lines, plants, enterprises, and supply chains

• Quickly accommodate and adjust for changes in demand and product

• Effectively manage product and manufacturing information.

1.7 VIRTUAL MANUFACTURING

The term Virtual Manufacturing is now widespread in literature but several definitions are attached to these words. First we have to define the objects that are studied. Virtual manufacturing concepts originate from machining operations and evolve in this manufacturing area. However one can now find a lot of applications in different fields such as casting, forging, sheet metalworking and robotics (mechanisms). The general idea one can find behind most definitions is that "Virtual Manufacturing is nothing but manufacturing in the computer". This short definition comprises two important notions: the process (manufacturing) and the environment (computer).

In [1.1, 1.2] VM is defined as "...manufacture of virtual products defined as an aggregation of computer-based information that ... provide a representation of the properties and behaviours of an actualized product".

18

Some researchers present VM with respect to virtual reality (VR). On one hand, in [1.3] VM is represented as a virtual world for manufacturing, on the other hand, one can consider virtual reality as a tool which offers visualization for DM [1.4].

The most comprehensive definition has been proposed by the Institute for Systems Research, University of Maryland, and discussed in [1.5, 1.6]. Virtual Manufacturing is defined as "an integrated, synthetic manufacturing environment exercised to enhance all levels of decision and control".

– Environment: supports the construction, provides tools, models, equipment, methodologies and organizational principles,

– Exercising: constructing and executing specific manufacturing simulations using the environment which can be composed of real and simulated objects, activities and processes,

– Enhance: increase the value, accuracy, validity,

– Levels: from product concept to disposal, from factory equipment to the enterprise and beyond, from material transformation to knowledge transformation,

– Decision: understand the impact of change (visualize, organize, identify alternatives).

A similar definition has been proposed in [1.7]: "Virtual Manufacturing is a system, in which the abstract prototypes of manufacturing objects, processes, activities, and principles evolve in a computer-based environment to enhance one or more attributes of the manufacturing process."

One can also define VM focusing on available methods and tools that allow a continuous, experimental depiction of production processes and equipment using digital models. Areas that are concerned are (i) product and process design, (ii) process and production planning, (iii) machine tools, robots and manufacturing system and virtual reality applications in manufacturing.

1.8 THE SCOPE OF VIRTUAL MANUFACTURING

The scope of VM can be to define the product, processes and resources within cost, weight, investment, timing and quality constraints in the context of the plant in a collaborative environment. Three paradigms are proposed in [1.5]:

a) Design-centred VM: provides manufacturing information to the designer during the design phase. In this case VM is the use of manufacturing-based simulations to optimize the design of product and processes for a specific manufacturing goal (quality, flexibility, …) or the use of simulations of processes to evaluate many production scenario at many levels of fidelity and scope to inform design and production decisions.

b) Production-centred VM: uses the simulation capability to modelize manufacturing processes with the purpose of allowing inexpensive, fast evaluation of many processing alternatives. From this point of view VM is the production based converse of Integrated Product Process Development (IPPD) which optimizes manufacturing processes and adds analytical production simulation to other integration and analysis technologies to allow high confidence validation of new processes and paradigms.

c) Control-centred VM: is the addition of simulations to control models and actual processes allowing for seamless simulation for optimization during the actual production cycle. Another vision is proposed by Marinov in [1.7]. The activities in manufacturing include design, material selection, planning, production, quality assurance, management, marketing, …. If the scope takes into account all these activities, we can consider this system as a Virtual Production System. A VM System includes only the part of the activities which leads to a change of the product attributes (geometrical or physical characteristics, mechanical properties, …) and/or processes attributes (quality, cost, agility, …). Then, the scope is viewed in two directions: a horizontal scope along the manufacturing cycle, which involves two phases, design and production phases, and a vertical scope across the enterprise hierarchy. Within the manufacturing cycle, the design includes the part and process design and, the production phase includes part production and assembly.

We choose to define the objectives, scope and the domains concerned by the Virtual Manufacturing thanks to the 3D matrix represented in Fig. 2 which has been proposed by IWB, Munich. The vertical plans represent the three main aspects of manufacturing today: Logistics, Productions and Assembly, which cover all aspects directly related to the manufacturing of industrial goods. The horizontal planes represent the different levels within the factory. At the lowest level (microscopic level), VM has to deal with unit operations, which include the behaviour and properties of material, the models of machine tool – cutting tool – workpiece-fixture system. These models are then encapsulated to become VM cells inheriting the characteristics of the lower level plus some extra characteristics from new objects such as a virtual robot. Finally, the macroscopic level (factory level) is derived from all relevant sub-systems.

The last axis deals with the methods we can use to achieve VM systems. These methods will be discussed in the next paragraph.

1.9 METHODS AND TOOLS USED IN VIRTUAL MANUFACTURING

Two main activities are at the core of VM. The first one is the "modelling activity" which includes determining what to model and the degree of abstraction that is needed. The second one is the ability to represent the model in a computer-based environment and to correlate to the response of the real system with a certain degree of accuracy and precision: the "simulation activity". Even if simulation tools often appears to be the core activity in VM, others research areas are relevant and necessary. One can find in [1.5] a classification of the technologies within the context of VM in 4 categories. A "Core" technology is a technology which is fundamental and critical to VM. The set of "Core" technologies represents what VM can do.

An "Enabling" technology is necessary to build a VM system. A "Show stopper" technology is one without which a VM system cannot be built and finally a "Common" technology is one that is widely used and is important to VM.

We propose the following activities to underline methods that are necessary to achieve a VM system:

− Manufacturing characterization: capture, measure and analyze the variables that influence material transformation during manufacturing (representation of product/process, design by features, system behaviour, …),

− Modeling and representation technologies: provide different kinds of models for representation, abstraction, standardization, multi-use, … All the technologies required to represent all the types of information associated with the design and fabrication of the products and the processes in such a way that the information can be shared between all software applications (Knowledge based systems, Object oriented, feature based models,…)

− Visualization, environment construction technologies: representation of information to the user in a way that is meaningful and easily comprehensible. It includes Virtual reality technologies, graphical user interfaces, multi context analysis and presentation,…

− Verification, validation and measurement: all the tools and methodologies needed to support the verification and validation of a virtual manufacturing system (metrics, decision tools,…).

− Multi discipline optimization: VM and simulation are usually no selfstanding research disciplines, they often are used in combination with "traditional" manufacturing research.

Numerous tools are nowadays available for simulating the different levels described in Figure 1.3: from the flow simulation thanks to discrete event simulation software to finite elements analysis. The results of these simulations enable companies to optimise key factors which directly affects the profitability of their manufactured products.

1.10 EXPECTED BENEFITS

As small modifications in manufacturing can have important effects in terms of cost and quality, Virtual Manufacturing will provide manufacturers with the confidence of knowing that they can deliver quality products to market on time and within the initial budget. The expected benefits of VM are:

– from the product point of view it will reduce time-to-market, reduce the number of physical prototype models and improve quality: in the design phase, VM adds manufacturing information in order to allow simulation of many manufacturing alternatives: one can optimize the design of product and processes for a specific goal (assembly, lean operations, ...) or evaluate many production scenarios at different levels of fidelity;

Manufacturing level	Type of simulation	Simulation targets	Level of detail
Factory / shop floor	- Flow simulation - Business process simulation	- Logistic and storage - Production principles - Production planning and control	low
Manufacturing systems / manufacturing lines	- Flow simulation	- System layout - Material flow - Control strategies - System capacity - Personnel planning	Intermediate
Manufacturing cell / Machine tool / robot	- Flow simulation - Graphical 3D kinematics simulation	- Cell layout - Programming - Collision test	High
Components	- Finite-Elements Analysis - Multibody simulation - Bloc simulation	- Structure (mechanical and thermal) - Electronic circuits - Non-linear movement dynamics	Complex
Manufacturing processes	- Finite-Elements analysis	- Cutting processes: surface properties, thermal effects, tool wear / life time, chip creation - Metal forming processes: formfill, material flow (sheet metal), stresses, cracks	Very complex

Figure 1.3. Manufacturing simulations.

– from the production point of view it will reduce material waste, reduce cost of tooling, improve the confidence in the process, lower manufacturing cost,...: in the production phase, VM optimizes manufacturing processes including the physics level and can add analytical production simulation to other integration and analysis technologies to allow high confidence validation of new processes or paradigms. In terms of control, VM can simulate the behaviour of the Manufacturing machine tool including the tool and part interaction (geometric and physical analysis), the NC controller (motion analysis, look-ahead)...

If we consider flow simulation, object-oriented discrete events simulations allow to efficiently model, experiment and analyze facility layout and process flow. They are an aid for the determination of optimal layout and the optimization of production lines in order to accommodate different order sizes and product mixes.

The existence of graphical-3D kinematics simulation are used for the design, evaluation and off-line programming of work-cells with the simulation of true controller of robot and allows mixed environment composed of virtual and real machines.

The finite element analysis tool is widespread and as a powerful engineering design tool it enables companies to simulate all kind of fabrication and to test them in a realistic manner. In combination with optimization tool, it can be used for decision-making. It allows reducing the number of prototypes as virtual prototype as cheaper than building physical models. It reduces the cost of tooling and improves the quality. VM and simulation change the procedure of product and process development. Prototyping will change to virtual prototyping so that the first real prototype will be nearly ready for production. This is intended to reduce time and cost for any industrial product. Virtual manufacturing will contribute to the following benefits [1.5]:

1. Quality: Design For Manufacturing and higher quality of the tools and work instructions available to support production;

2. Shorter cycle time: increase the ability to go directly into production without false starts;

3. Producibility: Optimise the design of the manufacturing system in coordination with the product design; first article production that is trouble-free, high quality, involves no reworks and meets requirements.

4. Flexibility: Execute product changeovers rapidly, mix production of different products, return to producing previously shelved products;

5. Responsiveness: respond to customer "what-ifs" about the impact of various funding profiles and delivery schedule with improved accuracy and timeless,

6. Customer relations: improved relations through the increased participation of the customer in the Integrated Product Process Development process.

1.11 AUTOMOTIVE

In the automotive industry, the objective of the Digital Product Creation Process is to design the product and the manufacturing process digitally with full visualization and simulation for the three domains: product, process and resources. The product domain covers the design of individual part of the vehicle (including all the data throughout the product life cycle), the process domain covers the detailed planning of the manufacturing process (from the assignment of resources and optimization of workflow to process simulation). Flow simulation of factories and ware houses, 3D-kinematics simulation of manufacturing systems and robots, simulation of assembly processes with models of human operators, and FEA of parts of the automobiles are state-of-the-art.

New trends are focusing on the application of Virtual and Augmented Reality technologies. Virtual Reality technologies, like e.g. stereoscopic visualization via CAVE and Powerwall, are standard in product design. New developments adapt these technologies to manufacturing issues, like painting with robots. Developments in Augmented Reality focus on co-operative telework, where developers located in distributed sites manipulate a virtual work piece, which is visualized by Head Mounted Displays.

1.12 AEROSPACE

Virtual Manufacturing in aerospace industry is used to design and optimise parts, e.g. reduce the weight of frames by integral construction, in 3D-kinematics simulation to program automatic riveting machines and some works dealing with augmented reality to support complex assembly and service tasks (the worker sees needed information within his glasses). The simulation of human tasks with manikins allows the definition of useful virtual environment for assembly, maintenance and training activities [1.10], [1.11].

1.13 CONCLUSIONS

As a conclusion of this chapter, we can say that we have now reached a point where everyone can use VM. It appears that VM will stimulate the need to design both for manufacturability and manufacturing efficiency.

Nowadays, even if there is a lot of work to do, all the pieces are in place for Virtual Manufacturing to become a standard tool for the design to manufacturing process: computer technology is widely used and accepted, the concept of virtual prototyping is widely accepted, companies need faster solutions for cost / time saving, for more accurate simulations, leading companies are already demonstrating the successful use of virtual manufacturing techniques.

Nevertheless, we have to note that there are still some drawbacks to overcome for a complete integration of VM techniques: data integrity, training, system integration. Moreover if large manufacturing enterprises have developed and applied with success VM technologies (aerospace, automotive, railway fields).

1.14 REFERENCES

[1.1] Iwata K., Onosato M., Teramoto K., Osaki S. A., Modeling and Simulation Architecture for Virtual Manufacturing System, Annals CIRP, 44, 1995, pp. 399-402

[1.2] Lee D.E., Hahn H.T., Generic Modular Operations for Virtual Manufacturing Process, Proceedings of DETC'97, ASME Design EngineeringTechnical Conferences, 1997.

[1.3] BowyerA., Bayliss G., Taylor R., Willis P., A virtual factory, International Journal of Shape Modeling, 2, N°4, 1996, pp215-226.

[1.4] Lin E., Minis I., Nau D.S., Regli W.C., The institute for System Research, CIM Lab, 25 March 1997, <www.isr.umd.edu/Labs/CIM/vm/vmproject.html>

[1.5] Virtual Manufacturing User Workshop, Lawrence Associates Inc., 12-13 July1994, Technical report.

[1.6] Saadoun M., Sandoval V., Virtual Manufacturing and its implication, Virtual reality and Prototyping, Laval, France, 1999.

[1.7] Marinov V., What Virtual Manufacturing is? Part I: Definition, 13 October 2000, <bosphorus.eng.emu.edu.tr/vmarinov/VM/VMdef.htm>

[1.8] Marinov V., What Virtual Manufacturing is? Part II: The Space of Virtual Manufacturing:, October 2000, <bosphorus.eng.emu.edu.tr/vmarinov/VM/VMspace.htm>

[1.9] Dépincé Ph., Tracht K., Chablat D., Woelk P.-O., Future Trends of the Machine Tool Industry, Technical Workshop on Virtual Manufacturing, October 2003, EMO'2003.

[1.10] Chedmail P., Chablat D., Le Roy Ch., " A distributed Approach for Access and Visibility Task with a Manikin and a robot in a Virtual Reality Environment ", IEEE Transactions on Industrial Electronics, august 2003.

[1.11] Di Gironimo G., Guida M.,"Developing a Virtual Training System in Aeronautical Industry", in Proceeding of the 5th Eurographics Italian Chapter, Trento, 2007.

## CHAPTER 2

VIRTUAL REALITY AND THE *VRTest* LABORATORY

2.1 VIRTUAL REALITY

Virtual Reality (VR) can be defined as technology that improves the interaction between human and product models, adding perception with realistic visual, tactile and sound sensations in a real-time simulated environment [2.1]. The user can experience the design phase of a new concept looking at it in the three-dimensional aspect, moving around or inside it, grasping parts, improving his analysis, comprehension, design and communication. The advantage of using VR in industrial applications is to reduce the time and costs of developing a new product, thanks to the new way of the design process: from the trial and error method, involving the production of many physical prototypes, to the continuous improvement approach, typical of simulation tasks [2.2]. The process of Virtual Prototyping uses digital models and VR techniques to simulate a product and its behaviour in a realistic way and to allow designers to test and evaluate their concepts. Many companies are already using VR techniques in the phase of Concept Design for styling decisions [2.3]: much less common is the application of VR for the simulation of maintenance tasks and maintainability tests, even if, for example, assembly-disassembly processes have a great affect on the product cost. Whoever manages the maintenance is interested in containing the costs caused by the frequency of maintenance operations and the necessary time and work for the execution of corrective and preventive activities. So it is necessary to optimize these operations by reducing the time and improving ergonomics, in order to limit the costs of staff training and postpone any operations not immediately necessary. A valid contribution to achieving these objectives can be obtained by simulating maintenance activities in a virtual environment [2.4].

New approaches are advancing the integration of the simulation of maintenance operations in the virtual design process in order to optimize the product in all the phases of its life [2.5]. Unlike traditional design processes, Virtual Assembly systems allow engineers to choose appropriate solutions, depending on the assembly sequences of digital models, that provide virtually tangible data even though without a real physical prototype [2.6]. Sequences and trajectories of assembly can be calculated, by means of collision detection in real time, in order to define the optimal path that guarantees the absence of interferences or penetrations among parts [2.7].

The same approach has been used to simulate manufacturing systems in order to create a digital copy of the work cells present in all production lines. The evolution of this research activity consists in the implementation of such simulations in a virtual environment allowing us to evaluate in real size active areas and safety margins along the production line. For example, in the study of workspaces respecting minimum and maximum distances between machines within the work cells has to be taken into account.

Virtual Reality is an indispensable instrument in making these analyses and simulations, visualizing results in an immersive environment, interacting with the models by means of the help of special devices for virtual navigation and handling, such as gloves, stereoscopic viewing and tracking systems. The direct manual interaction approach provides for the transfer of the movements of the user's body to the virtual scene in which the activity is simulated; in particular, a digital model of the hand reflects the position and orientation of the real hand in order to experience and test the task in the virtual environment. Different algorithms for virtual grasping have been studied in order to realize an even more realistic and comfortable simulation of the grabbing gesture [2.8]. To increase the realism of the operation attention has been paid to many aspects of the real world. Real time shadows and models of physical behaviours have been implemented in order to improve perception of the scene simulated in the virtual environment [2.9]. Moreover, Virtual Training is a natural application of such simulations, allowing the operators to acquire in advance experience with the operations they will execute on the real products.

2.2 THE VIRTUAL REALITY LABORATORY OF THE COMPETENCE TRANSPORT CENTRE: *VRTest*

In May 2005, a research group working at the Department of Progettazione and Gestione Industriale of the University of Naples Federico II completed the installation of a VR Laboratory, called "VRTest", realized for the Competence Regional Centre for the assessment of transportation systems set up by Campania Regional Authority. The laboratory is intended to be used for the advanced design of transport systems in the railway, aeronautical, naval and automotive fields. The plant, that is among the most important in Europe in terms of performance and dimensions, allows the designers to develop products and complex systems and simulate their configurations and performance in a virtual environment.

Figure 2.1. Visualization of an aircraft in full scale.

Therefore, it represents an ideal theatre for the immersion visualization in real size of important parts of manufactured components of great dimensions such as railway coaches or aeroplanes (v. fig.2.1). The *VRTest*, the technical characteristics of which are described in the following sections, is an integral part of the Competence Centre, related to Transport, with which Campania Regional Authority has been endowed thanks to important funding granted by the European Economic Community (POR Campania 2000-2006-MIS 3.16, deliberation C.R. of July 31 2001 n°s 3793), [2.10]. The Transport Competence Regional Centre (CRdC Trasport) has been realized with the aim of integrating the research of the Campania Regional Authority and of satisfying the innovation requirements in the transportation field. The project has provided a structure of research equipped with many technological laboratories able to integrate the competences of university and research institutions working in Campania, to service the technological development, the transport and the assessment systems of the local enterprises operating in the aeroplane, maritime and terrestrial field. Particularly, the project has identified three specific activity objectives: the reduction of environmental pollution, the safety of vehicles and the greater efficiency of transport systems. Hence the Centre assumes the acronym **Test** that stands for Technology, Environment, Safety and Transport.

VRTest has been designed in order to develop a role of horizontal integration and support to the activities of the other laboratories inside CRdC Transport. The strategic requirements defined in accordance with the objectives for the use of *VRTest*, make it a point of reference for those firms, working in the transport field, that would like to test in a virtual environment, with notable reductions of time and costs, aesthetic validations, verifications of assembly-disassembly and maintenance operations, production cycle simulations, kinematics and automated system

28

simulations, ergonomic analysis, conceptual design and virtual crash, as well as to give presentations to customers of virtual prototypes of new models which are imminently going to be put on the market. The choices of the technical characteristics of the laboratory derive from a careful analysis of the state of the art of Virtual Reality centres realized by European firms in the transport field. The laboratory consists of the following subsystems.

2.2.1 Graphic and Computational System

The graphic and computational unit has to allow for the visualization of images of great dimensions with extreme accuracy and fluidity of movement. These demands influence the characteristics of the computational system for the ability of geometry management, the complexity of the models and the treatment of the pixels in terms of image resolution. Therefore, the system has to be equipped with advanced graphics and significant computational potentialities to be able to perform at high speed the computational activities required for the simulation.

The problem in the choice of the graphic and computational system was closely connected to the choice of the visualization system; this was particularly true because we wanted to employ a projection system consisting of more than two projectors. The latest graphic cards of the Nvidia, in fact, allow the use of a single Windows/Linux workstation to perform a multi projection in active stereo on two projectors, with the possibility of covering, therefore, screens like Powerwall up to 5 meters long and managing automatically the functionality of Edge Blending. Obviously, this solution with a single workstation does not guarantee good performance when we have to manage complex models containing large quantities of polygons. In this case, and in the cases when we want to perform a multi projection with more than two projectors (for instance in the cases of Powerwall of greater dimensions than 5 metres long, of CAVE or of cylindrical screens) the problem is resolved, generally, either through the use of the Onyx SGI workstation of high performance with the Irix 6.5 operating system or through the use of a PC cluster with the Linux or Windows operating system .

For these reasons, and because the laboratory proposed to develop both an operational role of support to firms and research activities, the decision was taken to employ a mixed solution: SGI of the latest generation and elevated performance, the Onyx 4, and a cluster of three Windows/Linux workstations (Figure 2.2).

Table 2.1 illustrates the main characteristics of the SGI Onyx 4 workstation present in *VRTest.*

Table 2.1. SGI Onyx 4 Workstation at VR Test.

N. 10 Processori *MIPS R 16K* da 700 Mhz (4 Mb L3 Cache)
N.6 Graphic Pipe *InfinitePerformance* (combinate due a due mediante 3 Compositor per un totale di tre uscite principali verso la terna dei proiettori)

N. 3 *Compositor*
10 GB RAM: 10 banchi da 1 GB globalmente indirizzabili
N. 1 *SGI TP900 Base Unit Rack Mountable*
N. 2 dischi di sistema *Ultra SCSI3 Hot-swap*, 15K rpm, da 73 GB ciascuno
N. 8 dischi *Ultra SCSI3 Hot-swap*, 10K rpm ,da 146 GB ciascuno, inseriti nel *disk array TP900* collegato su porta SCSI LVD
Operative System IRIX Rev. 6.5

Figure 2.2. Graphic and calculus system at *VRTest*: SGI Onyx 4 (on the left) and Cluster of 3 Windows/Linux PCs (on the right).

Table 2.2 illustrates the main characteristics of the three PCs of the Cluster.

Table 2.2. PC Cluster characteristics at VRTest.

IBM INTELLISTATION ZPRO 6221
N. 2 Processori Intel Xeon da 3.06 Ghz
4 GB di Ram
Disco di sistema Ultra360 Scsi da 36 GB
N. 2 Dischi aggiuntivi Ultra360 Scsi da 73 GB
Graphics Nvidia Quadro FX3000G, 256 Mb Ram, 2 channels, support for stereoscopic view with shutter glasses.

2.2.2 Visualization System

The visualization system realized in *VRTest* is composed of a system of multi-channel video-projection with innovative high resolution DLP technology, that offers the advantage of greater brightness in comparison with the more traditional CRT technology, and a PowerWall screen for retro projection of great dimensions (7.50m x 2.40m). The image in full scale on the whole wall is achieved by placing side by side the multiple video projectors, that project, in parts, the general image onto the whole screen. The final scene has the characteristics of perfect continuity obtained thanks to the mechanisms of fusion of edge blending. The selected projectors are the

Barcos Galaxy 6 Classic + with DLP projection technology and native resolution of 1400x1050 pixels (v. fig. 2.3).

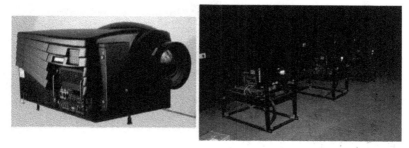

Figure 2.3. DLP Barco Galaxy 6 Classic + Video projector(on the left) Projectors Set up at VRTest (on the right).

They are equipped with 3 DMD type Dark Metal chips of Texas Instruments, and they are optimized for the stereoscopic vision in active formality. They reach a brightness of 6000 ANSI Lumen under nominal conditions, with a uniformity of non inferior illumination up to 90% and with a contrast of 1000:1. In the projectors some systems that allow us to improve their functionalities are integrated: the CLO (Constant Light Output) system guarantees that the brightness is constant and uniform in three channels; the OSEM (Optical Soft Edge Matching) system guarantees, instead, high and constant levels of contrast of optic blending; finally, the Barco WarpTM system allows a perfect geometric correction of the image. The lamps, from 1.2 Kws, are sufficiently characterized by a long duration (1200-1500 hours). These projectors introduce an extremely interesting denominated functionality, the-BlendTM. The selected screen is the denominated Barco CADWall ACTCAD (v. fig. 2.4), of dimensions of 7500 mm x 2400 mm.

2.2.3   Equipment For Stereoscopic View

When the active stereo projection is used, we need special liquid crystal glasses, called *shutter glasses*, and infra-red emitters that are installed above the screen. The stereoscopic view is given through the employment of synchronized shutter glasses with the update of the image produced by the installed graphic and computational unit.

Figure 2.4. Barco CADWall screen set up at *VRTest*.

The stereoscopic visualization is possible thanks to the simultaneous visualization of two images, one for each eye. The shutter glasses are equipped with, in place of the lenses, liquid crystals filters that, if correctly polarized, darken completely, preventing the eye from seeing through them. The images are shown in sequence on the screen, alternating, repeatedly, the frames for the right eye with those for the left eye. At the same time a signal is sent to the glasses in order to darken the filter of the eye not interested in the image at that moment on the screen. The advantage of this technology is represented by the high graphic definition, depending only on the system elaboration speed. In *VRTest* there are seven pairs of Crystaleyes-3 glasses (v. fig. 2.5) and three infra-red rays emitters installed above the screen.

2.2.4 Tracking System

The *VRTest* tracking system is the optical system with ART (Advanced Real-time Tracking) cameras. (v. fig. 2.6). It is characterized by high precision, immune from interference caused by metallic objects and/or magnetic fields and it is able to recognize and track 20 wireless sensors simultaneously.

Figure 2.5. Crystaleyes-3, for the stereoscopic view, available at *VRTest*.

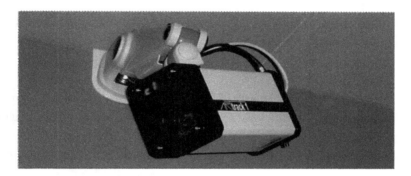

Figure 2.6. One of the three cameras of the optical tracking system ART at *VR Test*.

The three installed cameras above the screen allow us to cover an action area of 4m (width) x 3m (depth) x 2.5m (height). The widespread use of this system in the most important VR laboratories in the world has allowed it today to be directly supported by many VR software packages (for instance CATIA V5 by Dassault Systemes and Virtual Design 2 by vrcom), or through an interface software such as TrackD by VRCO (for instance Teamcenter Visualization 2005 by UGS and Amira VR by TGS-Mercury). Figure 2.7 shows some markers of the tracking system; they can be positioned on the hands, on the glasses or on any other object to be traced.

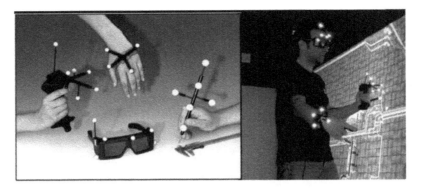

Figure 2.7. Some markers of the ART tracking system present at VRTest.

There are four essential components of the ART tracking system: the cameras (from a minimum of 2 to a maximum of 16), the software DTrack with the hardware, the flystick and the markers mounted on suitable supports for shutter glasses, hands, HMD helmets, stylus, flystick and others. These components are described in detail in the following sections.

The cameras

The cameras have infra-red sensors for CCD (Charge Coupled Device) images. An infra-red ray is periodically projected in the tracing area and when a marker is present it reflects that ray. The image acquired by the camera is sent through ethernet to the card mounted on the PC. The card transmits the data to the DTrack software that elaborates the data and calculates the position and the orientation. The last step consists in transferring the calculated data (by ethernet) to the applications that have to use it, in other words to VR software .

In the *VRTest* laboratory three cameras are mounted: one on the right, one in the middle and one on the left.

The DTrack software

The DTrack software checks every function of the whole tracking system and is equipped with simple programs in order to carry out different operations such as the calibration of the coordinate system of the room and the body calibration.

Flystick

The flystick (flying Joystick) is a wireless device, powered by a battery, designed for VR applications. It consists of a "cloche" with 8 buttons. The DTrack software draws the 6 degrees of freedom (DOF) of the object and is also able to recognize the pressure of every single button.

In Figure 2.18 the flystick is shown together with the base support (necessary for loading the battery) and the transceiver.

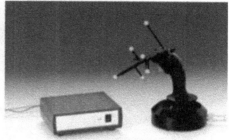

Figure 2.8. Flystick together with the base support and transceiver.

2.2.5 Audio System

The card audio system, integrated in the principal knot of the cluster of the PC, is "Creative Sound Blaster® Live! 24-bit"; it offers an elevated audio quality and high definition and it supports up to configuration 7.1 of the audio surround on the computer. Such a card allows us to use the EAX® technologies that, as has been said in the previous chapter, allow us to realize particular valid effects both for games and for listening to music but above all for VR applications. In *VRTest* five analogical loudspeakers such as subwoofers are installed.

2.2.6 3D Input Systems for navigation

In order to navigate in the virtual environment of *VRTest* it is possible to use both some devices from a console, such as Spaceball and Joystick, and the tracking system sensor for the tracing of the position of the user's head (Figure 2.9)

2.2.7 3D Input Systems for manipulation

For the manipulation of virtual objects in *VRTest* it is possible to use both the flystick, the position of which is traced in space by the ART tracking system, and the Cyberglove produced by Immersion Corporation (Figure 2.10). The movements of the hand in space are checked through one of the tracking system sensors.

Figure 2.9. Navigation by using Spaceball and Joystick from the console.

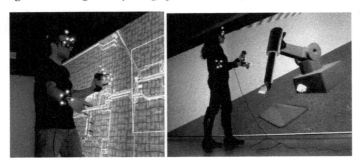

Figure 2.10. Manipulation by using Cyberglove and Flystick.

2.2.8 Room

The physical and visualization workspace is composed of a room, specifically designed to entertain the virtual immersive environment, providing suitable environmental conditions, with particular attention to aspects such as brightness, acoustics, air-conditioning and ergonomics. In order to perform such operations, the room is equipped with the following hardware tools.

Switching System and room control

In order to simplify the control and the management of the VR laboratory there are interconnection devices such as switching and serial control connection equipment, provided with touch screen control panels. The sources, coming from Onyx, the 3 PCs of the cluster, the PC that manages the tracking system, the VHS, the DVD or from a laptop computer, can be transmitted through a RGB matrix to a high passing band on the projectors and on the control monitors positioned on the console table (Figure 2.10).

Audio and Video Systems

The audio and video system is composed of a radio-microphone, an amplifier, a DVD, a VCR, 5 boxes and 1 subwoofer. The presence of 5+1 audio channels is necessary for those applications in which we want to reproduce realistically the sound direction too.

Console

In the *VRTest* laboratory there is a console with support for three 21" monitors and for the devices for the interconnection of the systems and the control of the room (Figure 2.10).

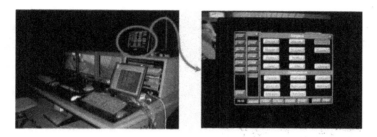

Figure 2.10. Console and Touch screen at *VRTest.*

2.2.9 Software

Since VRTest has the aim of developing several applications, it is equipped with CAD 3D modelling software and systems for the Digital Mock-Up analysis. Such analyses include: ergonomic verifications, assembly and disassembly of different components, maintainability analysis and style studies. Particularly, a significant part of the funding designated for the purchase of software was invested in one of the most important VR applications commercially available in Europe, Virtual Design 2 by vrcom. For the rest of the funding, the choice was made to purchase two complete PLM solutions offered by Dassault Systemes and by UGS respectively. Such solutions allow the realization of the majority of the analyses previously described. Particularly, the Dassault Systemes solution offers also an API environment (CAA-C++) and CAA-fine-workspace application builder, necessary when we want to develop applications in a virtual environment, not integrated generally in the system. Particularly, in *VRTest* the following CAD software is available: For-Engineering Wildfire 3 by PTC; Solid Edge V18 and Unigraphics NX by UGS; Catia V5 R17 by Dassault Systemes; Think Design 2006.2 by Think3; Alias Study Tools 13. The present virtual simulation software in *VRTest* is: Jack 5.0 (ergonomic analysis), Teamcenter Visualization 2005 (analysis of the Digital Mock Up, assembly/disassembly analysis and realistic visualization in an immersive environment) and Factory CAD 10 (simulation of digital factories) by UGS; Division Mock Up by PTC; Virtual

37

Design 2 by vrcom complete with Assembly, Showroom, Interior and Light Simulation modules and the development environment (see chapter 3).

2.3 ACKNOWLEDGEMENTS

The present work has been developed with the support of financial contributions from POR Campania 2000-2006 – MIS 3.16, provided to perform the activities of the Competence Centre for the Qualification of Transportation Systems founded by the Campania Regional Authority.

2.4 REFERENCES

[2.1] Caputo, F., Di Gironimo, G., Patalano, S., 2004, Specifications of a Virtual Reality centre for design in transports field (in Italian), Proceedings Of The XXXIII National Conference Adm – Aias Innovation In Industrial Design, Bari, 31 August – 2 September.

[2.2] Bonini, F., 2001, Numerical simulation and Virtual Reality Applied to the Real Word, Bollettino Del Cilea, 80: 34-35.

[2.3] Di Gironimo, G., Papa, S., 2006, Use Of Shader Technology For Realistic Presentation of Train Prototypes in Virtual Reality, Proceedings of the Fourth Conference of Eurographics Italian Chapter, Catania, 22 – 24 February: 105-109.

[2.4] Di Gironimo, G., Caputo F., Fiore, G., Patalano, S., 2004, Maintainability and Ergonomics Tests for the design of a Shunting Engine Through Virtual Manikins (in Italian), proceedings of the XXXIII National Conference Adm – Aias Innovation In Industrial Design, Bari, 31 August – 2 September.

[2.5] Gomes De Sá, A., Zachmann, G., 1999, Virtual Reality As A Tool For Verification Of Assembly And Maintenance Processes, Computers And Graphics, 23/3: 389-403.

[2.6] Li, J.R., Khoo, L.P., Tor, S.B., 2003, Desktop Virtual Reality For Maintenance Training: An Object Oriented Prototype System (V-Realism), Computers In Industry, 52: 109-125.

[2.7] Jayaram, S., Connacher, H., Lyons, K., 1997, Virtual Assembly Using Virtual Reality Techniques. In: Computer-Aided Design, 29/8: 575-584.

[2.8] Bruno, F., Luchi, M. L., Milite, A., Monacelli, G., Pina M., 2005, An Efficient Grasping Technique for Virtual Assembly Applications, Proceedings of XVII Ingegraf – XV Adm, Sevilla, 31 May – 2 June.

[2.9] Di Gironimo,G., Leoncini, P., 2005, Realistic Interaction for Maintainability Tests in Virtual Reality, Proceedings of Virtual Concept. Biarritz, 8-10 November.

[2.10]   Caputo F., Di Gironimo G., Patalano S., "Specifications of a Virtual Reality Centre for Transportation Design" (in Italian), Proceedings of the XXXIII National Conference Adm – Aias Innovation in Industrial Design, Bari, 31 August –2 September 2004.

CHAPTER 3

SIMULATION SYSTEM ARCHITECTURE

3.1   INTRODUCTION

Nowadays, the industrial scenario is characterized by the dynamics of increasing innovations and shorter product life cycles. Furthermore, products and their corresponding manufacturing processes are becoming more and more complex. Therefore, companies need new methods for the planning of manufacturing systems. To date, the efforts have focused on the creation of an integrated environment to design and manage the manufacturing process of a new product. The goal is to integrate Virtual Reality (VR) tools into the Product Lifecycle Management (PLM) of manufacturing industries. In order to realize this goal we have conducted a study to perform DM simulation steps and have provided a structured approach focusing on the interaction between simulation software and VR hardware tools to simulate both robotic and manual work cells.

This chapter is divided into four parts. The first part is a discussion of the action sequence to perform a manufacturing simulation. In the following section, DM and VM simulation system architecture is presented. Next, commercial software benchmarking is performed. In the last part of the chapter, a VR framework is described in detail.

3.2 ACTION SEQUENCE TO PERFORM A MANUFACTURING SIMULATION

In order to realize a virtual simulation in a manufacturing environment the following actions have to be performed (Figure 3.1):

1. Data collection (layout, 3D models, pictures, sequence of operations, assigned cycle time, productivity, availability, etc.). Prediction data from the Reliability & Maintenance (R&M) office are continuously collected and updated by monitoring and verifying in the field.

2a. Work cell simulation using dedicated software tools. Modularity of software is crucial in this phase to allow designers to work in parallel and to reduce work-time. Ergonomic evaluations for manual work cells and robotic simulations for automated work cells are included.

2b. Simulation of line/plant production flow . Until phase 2a. is completed, each machine or manual station can be represented as a "black box" with assigned cycle time, failure rate, availability etc.

3. Integration of each machine/station into discrete event plant simulation tool. Now the simulation is complete, and the output data can be analyzed.

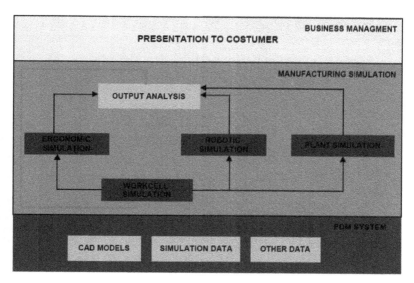

Figure 3.1. Sequence of actions to perform a manufacturing simulation.

4. Output analysis: it must be verified that each assigned parameter, such as technical efficiency, availability or throughput, fits with customer requirements (or rather with contract terms). If the data are satisfactory, the whole simulation work can be presented to the customer. If not, the data must flow back to the engineers (represented by a red line in the Figure 3.1 diagram) who can change the parameters and correct the problem.

5. Presentation of the whole work to the customer.

3.3 DIGITAL MANUFACTURING SIMULATION STEPS

To date, VR has been largely used to develop new methodologies for the design of automated assembly lines, in order to realize innovative products, the manufacturing process of which is automated as much as possible. The virtual simulation environment allows us either to evaluate the best workplace layout configuration, which minimizes the lead time in line production, or to optimize the automation level and the human component for each workplace. This means it is possible to carry out kinematic simulations of robots and ergonomic evaluations of operators in order to compare the results of these simulations with safety requirements. In manufacturing environments many software packages have been developed for virtual applications. These packages perform important functions that can be used to develop and create virtual manufacturing environments and to address process planning, cost estimation, factory layout, ergonomics, robotics, inspection, factory simulation and production management. Virtual Manufacturing can be realized by the integration of different software tools, each dedicated to

41

simulate three main production environments: robotized work cells, manual work cells and hybrid work cells [3.1], [3.2]. Figure 3.2 represents a DM architecture in which the simulation model is carried out by using four kinds of software.

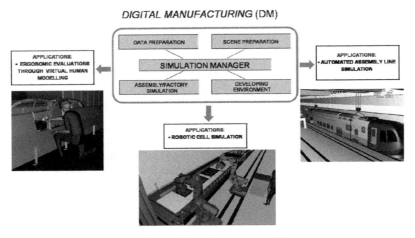

Figure 3.2. Digital Manufacturing Architecture.

A CAD package is used for modelling the geometry of components. Three simulation tools, classified for the work area, such as discrete event simulations for plant simulation, robotics and ergonomics, are used to simulate each workplace and the overall flow of the line. Table 3.1 shows the most important DM software classified for the work area adopted in manufacturing fields.

Table 3.1. Digital Factory software

TOOL	SOFTWARE HOUSE	AREA
ARENA	ROCKWELL	**DISCRETE EVENT PLANT SIMULATION**
AUTOMOD	BROOKS AUT.	
DELMIA - QUEST	DASSAULT SYSTEMES	
FACTORY CAD	UGS	
DELMIA - IGRIP	DASSAULT SYSTEMES	**ROBOTICS**
eM-WORKPLACE (ROBCAD)	UGS	
DELMIA - ERGO/HUMAN	DASSAULT SYSTEMES	**ERGONOMIC EVALUATIONS**
JACK	UGS	

3.3.1 Object modelling

The CAD system used for modelling objects is both a two and three-dimensional (2D-3D) design and drafting platform that automates design tasks. The components and sub-assemblies of the product and tools, the input and output bins and the sub-assemblies and fixtures are modelled manually using the CAD tools. The models of the components and the assemblies are exported into the simulation software using appropriate file export features.

3.3.2 Robotic simulations

In modern production plants, it is easy to observe an extensive use of robots, principally anthropomorphous robots. They can render fast, accurate and efficient a wide range of operation such as various kinds of assembly, "pick and place" operations, fastening and welding. In general, Computer Aided Robotics Systems (CAR-Systems) are used to design robot cells and to create the offline programs necessary to reduce start-up time and to achieve a considerable degree of planning reliability. The use of simulations for the planning and designing of robot cells and plants has increased substantially in recent years. Generally, commercial CAR-Systems can be divided into two main groups: high-end and low-cost. The best known high-end CAR-Systems used in the manufacturing environment are eM-Workplace (formerly ROBCAD) by UGS and Delmia - IGRIP by Dassault Systemes. These systems have been developed to meet the specific requirements of the production industry and can accomplish an amazing array of tasks. They integrate functionality for special tasks such as spot welding, laser application dispensing, and painting. With these tools, it is possible to simulate multiple robots from different manufacturing processes at the same time.

A comparison between eM-Workplace and IGRIP is shown in Table 3.2, focusing the attention on features such as CAD supported/data exchange capability, integration with VR tools and open architecture.

Table 3.2. eM-Workplace vs. IGRIP

MAIN PROPERTIES	SOFTWARE	
	IGRIP – DELMIA	EM-Workplace (formerly ROBCAD) - UGS
CAD SUPPORTED/DATA EXCHANGE CAPABILITY	CATIA, UNIGRAPHICS, PRO/ENGINEER, CADSS/NEUTRAL TRANSLATOR INCLUDE IGES,DXF,STEP,STL,DWG	CATIA, UNIGRAPHICS, PRO/ENGINEER/DIRECT CAD INTERFACE OR NEUTRAL FORMATS SUCH AS IGES, DXF, STL, STEP, ROBFACE- A TECNOMATIX NEUTRAL DATA EXCHANGE FORMAT ENABLES THE IMPLEMENTATION OF ANY KIND OF DEDICATED MCAD INTERFACE
INTEGRATION VIRTUAL REALITY TOOLS	CYBER-GLOVE FROM VIRTUAL TECHNOLOGY, ASCENSION TECHNOLOGY CORPORATION FLOCK OF BIRDS, STEREO DISPLAY INTERFACE	VD2 (VRCOM), INVISION (INTRO)
OPEN ARCHITECTURE	ALLOWS USERS TO PROGRAM CUSTOM FUNCTION WITH UNPARALLELED EASE BY CREATING MENU FUNCTIONS, CUSTOM DEVICE KINEMATICS AND MOTION PLANNING ALGORITHMS	THE TOOL EM-ROSE API OFFERS AN OPEN SYSTEM ENVIRONMENT FOR DEVELOPING CUSTOMIZED FEATURES AND APPLICATIONS

3.3.3 Ergonomic simulations

In the eighties and early nineties manufacturing industry exponentially increased the automation level in factories. OEMs were dreaming about a completely robotized production system. In the course of time, however, this ideology collapsed because of many obstacles and contra-indications, such as low availability and efficiency and the high cost of machines and maintenance. Nevertheless, the principal reason, in the recent past, that has re-introduced an extensive use of manual operations in factories, has been the irruption into the world scenario of new markets (e.g. the Far East and South America), with dramatic manpower cost reductions. For this reason ergonomic simulation is getting more and more critical in a Virtual Factory approach. Two of the main commercial software tools for ergonomic simulation are analyzed in this study, Delmia – ERGO (by Dassault Systemes) and JACK (by UGS) [3.3] It must be made clear that ERGO is more properly an add-on of IGRIP; in particular it can be used to model assembly and materials handling operations between workstations. This tool is basically used to design safe working environments that accommodate a wide range of workers and for ergonomic assessment and task analysis. It is used to address the human interface issue that impacts on the ability of a wide range of humans to assemble the prototype product and the process times needed for each task. Libraries of whole body, head, arms and hand postures are used. The software also provides "point and click" routines to generate walking, climbing, lifting and carrying sequences. In ERGO, to model a workstation operation, the worktable, parts and bins and human operator are imported as "devices" and placed at appropriate locations in the workstation. Using the ergonomics option, the human device was "taught" to perform the assembly process by creating a series of positions of the human hands while holding and assembling the product [3.4]. Figure 3.3 shows examples of workstations and operators modelled in IGRIP/ERGO, while Figure 3.4 shows a work cell in a train assembly cycle realized by JACK.

Figure 3.3. Work cell simulation with IGRIP (on the left) and ERGO (on the right)

Figure 3.4. Simulation of a Manufacturing Process with JACK: working on a work cell in a train assembly cycle .

3.3.4 Plant simulations

In accordance with the steps detailed in section 3, discrete-event plant simulation tools are also analyzed. ARENA is a 2D plant workflow simulation. It is more oriented to output data (e.g. reliability, productivity, efficiency and bottleneck detection) than to a graphic representation of the plant itself. AUTOMOD, on the other hand, allows a three-dimensional visualization [3.5]. Delmia - QUEST is an object-based, discrete event simulation tool. It is used to model, experiment with, and analyze facility layout and process flow. It provides visualization and data import/export capabilities.

The process of incorporating workstation sub-models into the discrete-event simulation model is basically simple. Initially, the workstation operation is played and recorded in the graphical modelling application. It is then exported as a QUESTCELL file. The display option provides, in the discrete event-simulation tool, the display of each station as the imported workstation sub-model. Similarly, the scripts associated with a human carrying a bin of components or subassemblies between stations are determined and imported. Figure 3.5 shows what the QUEST simulation model looks like after importing and integrating the workstation models that were developed in IGRIP [3.6].

A concurrent discrete event tool is FactoryCAD by UGS. This software allows us to work with "smart objects" that represent virtually all the resources used in a factory, including floor and overhead conveyors, warehouses and cranes and material handling containers and operators. With these objects, it is possible to "snap" together a layout model without wasting time drawing the equipment. Starting from a 2D layout drawn using Autocad GUI, FactoryCAD allows us to develop the project in 3D using a complete 3D object library.

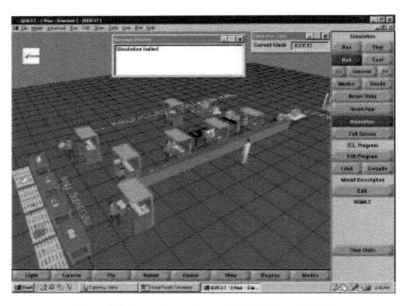

Figure 3.5. Plant simulation with DELMIA - QUEST.

46

3.4 VIRTUAL REALITY FOR MANUFACTURING SYSTEM SIMULATION

Taking into account what has been stated in Section 3.1, we have investigated the possibility of integrating VR tools in the overall flow of a vehicle production simulation (Figure 3.6). First of all, in order to achieve this integration, it is necessary to develop a procedure for Virtual Prototyping that provides, in particular, an optimization of the management of data exchange protocols between 3D parametric CAD systems and visualization environments using meshed models. One of the most pressing issues facing industry is data integration. The CAD systems used to realize product models are generally not suitable for producing a representation conducive to large scale and the frame rate guaranteed visualization required by VR applications. Although addressed to some degree by commercial providers of visualization software (such as UGS PLM Solution and Dassault Systemes) there is no general non-proprietary way to convert a CAD assembly into a representation suitable for VR (Figure 3.7).

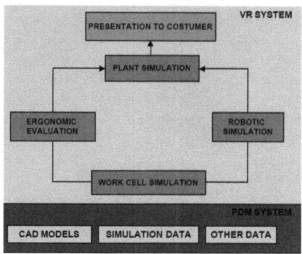

Figure 3.6. VR integration in overall simulation flow

The second step is the arrangement of a virtual environment in which it is possible to simulate assembling operations of products belonging to the manufacturing field, which means building up the "scene" where the operations take place. Then it is possible to realize the work cell simulation by integrating VR tools. As Figure 3.6 shows, there are three fields in which VR can be integrated: ergonomics, robotics and factory layout simulation.

47

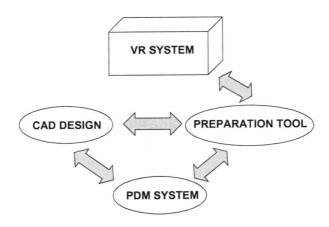

Figure 3.7. Data flow between CAD and VR system.

3.5 VIRTUAL MANUFACTURING ARCHITECTURE

Starting from DM architecture (Figure 3.2) we arrive at a VM simulation architecture introducing the immersion and the human interaction (Figure 3.8).

In such a way we add the direct human interaction with the digital mock-up to the applications performed in the DM architecture system simulation and therefore the possibility for the user to be immersed in the same simulation environment.

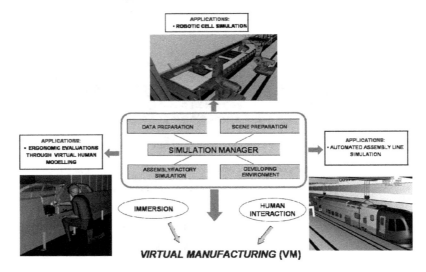

Figure 3.8. From Digital Manufacturing to Virtual Manufacturing Architecture.

The realization, in an immersive virtual environment, of manufacturing system simulations, introduced above, needs specific requirements:

- a powerful graphic and calculus system able to manage a great amount of data;
- a large screen able to visualize complex systems in full scale;
- input devices allowing the protagonist of the virtual experience to navigate easily and interact with the virtual scene and other members of the design team to share such an experience and review the design;
- software tools for collision detection, motion programming, kinematics simulation, 3D distance measurements, virtual mark-up and path recording;
- 3D audio output devices increasing the immersion in the virtual environment.

In order to satisfy these requirements an original VR architecture has been conceived (Figure 3.9). The architecture is based on complex hardware and software technologies available at the VR Laboratory, *VRTest*, the characteristics of which have been described in chapter 2.

The first requirement is widely satisfied by means of a graphic and calculus system characterized by high performance and flexibility. The visualization of the scene is obtained by means of a semi-immersive system composed of a power-wall, three DLP projectors and shutter glasses for active stereoscopic view. The main advantage of having a VR laboratory able to visualize industrial products and work cells that present large dimensions is the carrying out of the analyses in the virtual environment in full scale.

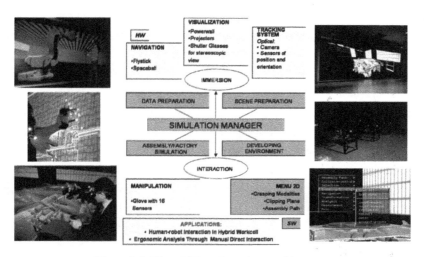

Figure 3.9. Virtual Manufacturing Architecture.

49

The platform used as the Simulation Manager is Virtual Design 2 (by vrcom): it is a comprehensive tool containing many functions for product development, from the creation of the virtual environment to assembly simulation or ergonomics analysis. Since VR software does not offer, to date, standard characteristics to perform engineering applications, in particular maintainability tests and manufacturing systems simulations, VD2 was chosen as it provides Application Programming Interfaces used to customize the architecture Simulation Manager.

This platform gives the possibility of interfacing with a wide range of input/output devices, such as those available in *VRTest*. The software allows us to manage all the input and output devices employed for the visualization, navigation and interaction with the scene. Due to this chosen approach, based on the direct manual interaction, the devices implemented in VD2, in particular for navigation and manipulation, have been:

- a flystick for the activation of actions by means of the events associated with the eight keys;
- a system of three cameras for tracking the position and orientation of the sensors attached to the head, hand and other characteristic human body points, in order to transfer real movements to the virtual scene;
- a 5DT glove with 16 sensors for the realistic movement of the fingers of the hand to interact with the geometries of the scene.

The navigation in the scene has been simulated by means of sensors of the tracking system attached to the glove controlled with one button of the flystick: the user protagonist of the virtual experience can move the point of view in the scene, clicking the assigned button of the flystick in the left hand and simultaneously dragging the scene with the movement of the hand. This user-friendly modality of navigation allows the operator to decide easily when to use his hand to change the position in the scene, for example for better positioning the geometries during the maintenance operation, rather than to execute the task. This modality of navigation allows also the other members of the design team to participate in the virtual immersive experience in a convenient way since they are not obliged to hold continuously the point of view of the protagonist of the virtual experience. Moreover, a sensor can be attached to the glasses of the protagonist and assigned to the camera control of the software in order to generate relative variations of the point of view by head rotations. In this way, for example, immediate inspection of the analyzed system is possible for the user simply by moving his head in a natural gesture.

The virtual manipulation of the objects has been implemented through a glove provided with sensors able to detect the position and movement of the real fingers sending data to the corresponding hand in the virtual scene. Three grasping modalities have been implemented: index-thumb grasp, palm grasp and gesture recognition grasp. The system allows us to choose between the palm, the simultaneous collision with index finger and thumb, and the recognition of a specific gesture (this last modality can be opportunely programmed).

The first operation for creating the virtual environment is to import all geometries necessary to carry out the simulation: the formats used for data exchange are VRML and 3ds and then imported files are optimized in order to be treated with the VR software. Once the geometries are imported it is possible to set the objects that will act as tools in the maintenance tasks, in order to define automatically the collisions with the virtual hand and with all the other "interactive" geometries. There are three modes to set the interaction of the geometries in the virtual environment: static, grabbable and glideable.

A component defined static cannot be grasped and moved with the hand, but is subject to collision detection with the other parts defined interactive. The alternatives, grabbable and glideable, concur to make it possible for an object to be grasped, but in different modalities. A component defined grabbable can be attached and moved by the hand. With glideable it is possible to produce a realistic behaviour of such parts that can slide along the generatrixes of coincident surfaces between two parts, remaining consistent with the hand. Moreover, it is also possible to introduce some cinematic restrictions in order to respect the real constraints of some components, and so facilitate the simulation, removing the less important movements. The simplest example is the sliding of a screw along its axis. It is also possible to assign a common axis to objects and tools so that, in accordance with a predefined distance, the instrument snaps to the correct position and the operator is guided in the task; otherwise it would be impossible in the simulation to realize the correct relative positioning of the axes with the current VR devices. Once the coincidence between the axes of tool and part have happened, it is possible to visualize the value of the rotation angle of the tool, in order to study for example accessibility without collisions during the task.

Different models of the virtual hand can be chosen in order to simulate the verification activities with hands characterized by different anthropometric dimensions relative to the different population percentiles. Moreover, another function can be activated: a 'ghost' mode can be set to visualize correctly a component fixed in the last position free from interferences, while the ghost object follows the hand up to the successive collision free position. An interactive 2D menu containing many functions that the customer can recall directly during the simulation by means of predefined events has been implemented (Figure 3.8). It is possible to select in the menu those objects that have to become grabbable to execute the current task.

Moreover, it is possible to record, load, save or modify the path of a virtual component while an assembly operation or a manufacturing process is performed. The path is represented by means of the polylines interpolating the positions occupied by the centre of gravity of the selected object. A clipping plane can be set from the 2D menu in order to visualize the inside of the assembly by means of sections at predefined distances.

3.6   VIRTUAL DESIGN 2 FRAMEWORK

The VR kernel for Virtual Design 2 consists of three main components [3.7]:

- **Interaction Manager**, controls all actions in the virtual environment as well as the user's interactions with the virtual scene.
- **Device Manager**, initializes and controls the hardware devices used in the virtual environment. It provides the mapping of physical devices to logical devices.
- **Rendering Kernel**, based on OpenGL, loads the geometry, maintains a hierarchical scene graph and renders it. The rendering kernel maintains a hierarchical scene graph with assembly-type nodes as branches and geometry-type nodes as leaves. See Figure 3.10.

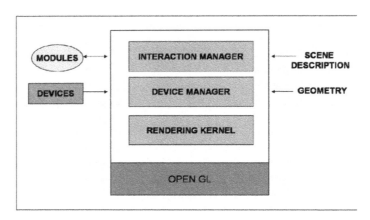

Figure 3.10. VD2 kernel components.

3.6.1 The Interaction Manager

There are basically two approaches to writing virtual environments (VE):

- Event based:

This approach begins with the creation of a story-board, i.e., the creator specifies which action/interaction happens at a certain time, because of user input, or any other event. A story-driven world usually has several "phases", so certain interaction options should only be made available at that stage of the application where it is appropriate, and others at their appropriate stage.

- Behaviour based:

52

Another approach is to specify a set of autonomous objects or agents, that are equipped with receptors and react to certain inputs to those receptors. So, overstating it a little, we take a bunch of "creatures", throw them into our world, and see what happens.

Since Virtual Design 2 is targeted at manufacturing industries, Virtual Design 2 pursues the event-driven approach. Of course, it is easy to write plug-ins which will implement behaviour in autonomous objects.

The Virtual Design 2 concept is to provide the basic interaction functionality frequently needed in virtual prototyping applications and related fields. We will describe the inner workings of some of this functionality below. It can be configured with a single scene description file, a script file which describes the static and dynamic configurations. The description is based on 'events', 'actions' and 'objects'. The basic idea is that certain events will trigger certain actions, properties, or behaviour. For example, when the user touches a virtual button, a light will be switched on or, when a certain time has been reached, an object will begin to move. The action can be specified in a configuration file

The basic action set can be functionally enhanced by dynamically loaded modules.

When these are used, they must be specified in the scene description file and then loaded by the interaction manager.

Collision detection

For collision detection we intend the mechanism that allows us to verify a collision between two or more elements of the virtual scene. It deals with one essential characteristic all interactive experience: for instance the possibility to manipulate virtual objects (grabbing) is subordinated to the individualization and the signalling of the contact among virtual hand and object to be manipulated. The main assignment of the collision detection module (CDM) is to produce an event "collision" every time there is a condition of intersection among two elements of the scene held under control and to allow interaction manager to undertake the action previously prepared. A collision detection module of VD2 consists of two main parts: a *front end* and a *back end*. The front end handles all requests, keeps track of "collidable" objects, and handles all underlying sub-modules (see figure 8.1). The backend is instead the module that notices the collision among two specified polyhedrons. Since the assignment of the back-end is computationally onerous, (independently by the used algorithm), the CDM do not keeps trace of every collision that happen among the objects in the virtual scene; it could be even unwanted for the industrial applications, in which we are generally interested in to manage only a limited subset of the possible interactions. For this reason it is necessary the programmer declares to the CDM not only what the objects are for which the control should be activate (collidable objects), but also what the couples of objects are whose collision have to be indeed signalled (collisions of interest). The collision detection algorithm can be considered as a pipeline of filters in succession. At every iteration the render Y signals in fact to the CDM the list of the objects that have been moved in comparison to the preceding frame.

The front-end module filters only from the list the objects declared collidable, that will be then compared with the matrix of the collisions of interest. In this way it will be possible to verify only the couples of interest in which at least one of the objects has changed position. At this point the back-end assignment is to determine the presence or not of the collision among the different couples.

3.6.2 The Device Manager

Device Manager, initializes and controls the hardware devices used in the virtual environment. It provides the mapping of physical devices to logical devices.

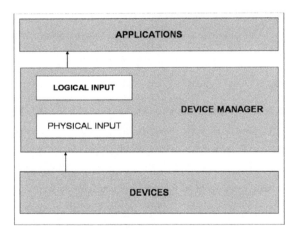

Figure 3.11. The Device Manager.

This concept simplifies scene building and enhances portability because one does not have to be concerned about the availability of hardware, such as which tracker-system is used or which machine or port it is attached to. A number of new interaction devices have been developed over the past few years to simplify interaction between CAD systems and immersive virtual environments.

These devices contain sensors for positioning, orientation in space, bending finger joints, etc. and are often referred to as "multi-dimensional interactive" (MDI) devices. They are connected to a computer, hereafter referred to as "host", via a serial interface. The host must communicate with each of the attached devices in the device manufacturer's proprietary protocol. The host for these devices is not necessarily the computer on which the actual application is running. Often the machine where the application is running has no free ports for attaching these additional devices or the application is running on a workstation that is quite remote from the workplace. In such cases, the MDI device host must transmit the data from these devices to the computer on which the application is running via a network connection.

IDEAL has been given a client-server architecture suitable to these types of hardware configurations. A so-called "device server" process, often simply called "server", runs on the host for each active MDI device. It is the application itself which plays the role of client when it communicates with the severs over the local network,. It is not necessary for the application to know just which specific device is actually being used or exactly where it is attached. For the application, it is only important to know what the nature of data is that the device delivers (in the sense of 3D positions or the finger positions of a hand).

There are many sources for error in a complex system. A device can be erroneously attached or configured, the cable or device itself may be defective, etc. In practice, such errors are more the rule and less the exception. A system such as IDEAL must be prepared to react accordingly. An error may not be allowed to cause the application to fail, or indeed to terminate. IDEAL will typically halt the application and report the error in a dialog box. This gives the user an opportunity to fix the problem and resume his work with the application. Each device server also produces a log of all significant actions and errors that occur in a file. This log can be quite helpful in troubleshooting errors afterwards.

There are many different interaction devices available on the market. The devices which are supported by IDEAL are briefly described below. Use of these devices with Virtual Design 2 will be treated in more detail in another section further back in this guide.

Multi-dimensional interactive devices can be divided into two classes:

- Devices placed on a flat surface and typically used in conjunction with a workstation form one class. An example of such a device is the spacemouse. Since these devices are stationary, they tend to deliver relative data, i.e. usually speeds rather than absolute positions.

- Devices attached to one's body or clothing, such as the Cyberglove, form the second device class. These devices are usually used for immersive applications. They deliver absolute data, i.e. the angle of bend for a body joint or the position of a tracker within a room.

55

The separation of these two classes cannot be taken too strictly as, occasionally one finds a Cyber glove at a workstation whereas sometimes a spacemouse will be used for immersive environments too, like in a CAVE for navigation.

As mentioned before, IDEAL has a client-server architecture. IDEAL will start a server process for every active device. These servers manage their respective devices and pass device data to the application. Virtual Design 2 therefore represents the client. In addition to application and server processes, a third group of processes play an important role for IDEAL, the demon processes.

A demon process must run on every computer to which an interactive device is attached, i.e. every host. The objective of these demon processes is, as with other UNIX demon processes (like send mail or dns), to provide a service.

In the case of the IDEAL demon (hereafter only referred to as 'demon'), this service consists of starting, monitoring, and terminating device servers. The demon must therefore be running on the host before a device can be started.

The most reliable manner to guarantee this sequence is to start the demon automatically when UNIX is booted. Your system administrator is responsible for this. Additionally each VD2 start script will test whether a demon process is running and start a new one if necessary. Then, if a device attached to a host is used by the application, IDEAL will first build the so-called "socket" connection to the demon on the affected host and request that it start a suitable server for the specified device.

The demon will then read a configuration file, called 'ideal.server' or Device File, and search it for the entry that corresponds to the specified device. If the entry is found, then the appropriate server process can be started. This server will initialize the attached device and builds its end of the socket connection to the application. Device data passed to the application and parameters received from the application must be moved through this socket connection.

3.6.3 The Rendering Kernel

The rendering kernel (called 'Y'), The basic structure for storing geometric information is different than that other commercially available systems like Inventor and Performer use. Those software packages differentiate point attributes (position, color, texture coordinates) which are stored in separate arrays and are allocated only as needed. Virtual Design 2 stores the points of a face as described by indices into an array. Each point of the rendering kernel is a structure which stores all point attributes at once. The points of a face are indices in the point array. Each face can have its own material and appearance (e.g. wire-frame, solid). An object is built from a list of faces and a list of points. This design has several advantages. All attributes of a point are

56

easily accessible. Furthermore, faces can be accessed directly and its attributes changed independently. This is very important for collision detection to show colliding faces. For instance, in order to achieve that behaviour with Performer, it is necessary to maintain 2 face lists; one for solid faces and one for wire-frame faces. To switch a single face from solid to wire-frame, that face has to be deleted from the 'solid' face list and inserted into the 'wire-frame' list. The main disadvantage to this approach is that memory is used more extensively, particularly if only point positions are needed. If this is a major problem for an application, it should be easy to add another node type to the system that is based on separate arrays or on smaller point-structures. The rendering kernel supports multiple graphics pipes and more than one rendering window per graphic pipe. The rendering kernel also offers several built-in functions for stereo viewing. Stereo viewing can be achieved with dual pipe rendering, shutter, MCO-style, interlaced, or anaglyph (green-red stereo).

Furthermore, CAVE projection is implemented with continuous lighting on all CAVE walls. The rendering kernel can use multiple processes for culling and drawing. It Is also capable of loading and saving Inventor/VRML geometry files.

Aside from the creation and manipulation of objects, nodes and materials, higher level functions are also embedded in the rendering kernel. For example, the rendering kernel supports back-to-front sorting for objects and faces to support correct transparency.

3.7   VIRTUAL ENVIRONMENT PROGRAMMING

In order to be usable by the final user (designer, experts of ergonomics, staff assigned to the maintenance), the VE must be opportunely programmed: such activity consists of previously defining the behaviour of each element of the scene in reply to the interaction with the user. Programming phase of the VE (authoring) is, therefore, fundamental: the more accurate such phase is, the more realistic the behaviour of the environment will result and consequently the reliability of the outcomes reached with the virtual simulation will be. The advantage of the chosen platform is that the authoring activity can easy developed not only by expert programmer, but also by CAD operator or simple user: such activity in VD2 is possible by means the writing of scripts, in the shape event trigger action, enough simple to assimilate (figure 6.1). Moreover, as said before, the software offers the possibility, thanks to the open developing environment, to write any plug in, in a common language, such as C or C++, which, once compiled in a shared library (DSO), can be called at the happen of defined events (e.g. a collision, the pressure of a key, a particular gesture of the virtual hand) as well as any other base command. Such possibility gives practically infinite possibilities to programme the Virtual Environment. It is possible, for example, to programme the reproduction of a sound (action) at the happen of a defined condition of collision between two elements of the virtual scene (input). In general, inputs are produced by physical devices, such as gloves or tracking systems, but they can be generated by the same software too, such as in the case of the collision detection. The actions on the objects of the scene can be represented by both base or external libraries.

3.8   DYNAMIC SHARED OBJECTS (DSO)

As aforementioned, the features provided by the VD2 kernel can be enhanced by functions defined in external modules, called Dynamic Shared Objects (DSO). Generally, each DSO module contains a set of functions developed for a specific application target, such as a plug-in. The basic installation of Virtual Design 2 already provides many plug-ins, for instance to manage interactive menus or to make snapshots of the virtual scene. A DSO module is a dynamically linkable object file, which allows the linking of the module to the VD2 kernel to be made at run-time, [3.8]. Moreover, the module is shared, meaning that many different processes can share the library functions at the same time, see Figure 3.12.

This modular approach offers three main benefits:

1. The object code is loaded in the physical memory only once and then it can be used by multiple processes via virtual memory management, [3.9];

2. It is easy to add new features to Virtual Design 2 and maintain them;

3. The object code is linked to the VD2 kernel only when the features implemented in the module are really needed. The functions provided by DSO modules can be used as well as the basic commands, by specifying them in the scene description file.

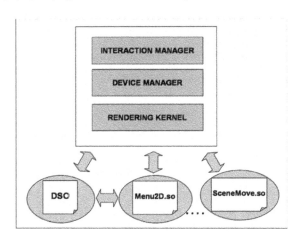

Figure 3.12. DSO modular approach.

58

3.9 CONCLUSIONS

The present work wants to demonstrate the important potentialities offered from VR techniques in industrial applications, in particular, for manufacturing system simulation. Obtained results not only provide a valid answer to the design questions in the field of manufacturing system simulation, but they make objective the effective applicability of the proposed methodology: in spite of the subjective character of the approach to the simulation, based on the direct manual interaction, the information collected in the case studies described in the following chapters, allow us to grant the feasibility of tasks and to individualize the design parameters on which operating to better answer the functional requirements and, finally, to improve the design. After each design modification, a new phase of simulation follows in order to verify the effective satisfaction of the requirements. In particular, the virtual simulation is important also for the training of the staff assigned to the maintenance and manufacturing activities.

The analyzed case studies have highlighted that the proposed architecture satisfies all the requirements needed to perform manufacturing system simulations. Nevertheless future works have to be focused on the implementation of more grasping conditions of the virtual objects, in order to increase the realism of the simulation, reproducing the natural posture of the hand assumed in the manipulation of the objects. Also the possibility to introduce a virtual reproduction of the wrist, the forearm and the arm of the operator, represents an interesting objective, in order to increase the sense of immersion into the scene and to make more precise and concrete the results of the simulation. Moreover, a future interesting task will be to implement algorithms for the automatic calculation of assembly/disassembly paths, starting from the initial and final position of the components: obtaining the theoretical collision free path, the simulation can allow the operator to reproduce the previously calculated trajectory. Finally the introduction of shadow effects on the examined system would increase the realism of the scene, in order to improve the perception of distances and volumes between virtual hand, tools and components.

3.10 REFERENCES

[3.1] Caputo F., Di Gironimo G., Marzano A., "A Structured Approach to Simulate Manufacturing Systems in Virtual Environment", XVIII Congreso International de Ingegneria Grafica, 31st May – 2nd June, Barcelona, Spain, 2006.

[3.2] Di Gironimo G., Marzano A., Papa S., "Design of a Virtual Reality Environment for Maintainability Tests and Manufacturing Systems Simulations", Proc. of 5th CIRP International

Conference on Intelligent Computation in Manufacturing Engineering, Ischia (Naples), Italy, 26th – 28th July 2006.

[3.3] DI GIRONIMO G.: Studio e Sviluppo di Metodologie di progettazione ergonomica in ambiente virtuale, Ph.D Dissertation, University of Naples, 2002.

[3.4] DONALD D. L.: A Tutorial on Ergonomic and Process Modeling Using QUEST and IGRIP. In Proceeding of the Winter Simulation Conference IEEE, Washington, D.C., U.S.A., December 13th - 16th ,1998: vol.1, 297-302

[3.5] ROHER Matthew W., Mc GREGOR Ian W., Simulation Reality using AUTOMOD. In Proceeding of the Winter Simulation IEEE, San Diego, CA, U.S.A., December 8th -11th ,2002, vol.1,173-181

[3.6] LAUGHERY R.: Using Discrete-Event Simulation to Model Human Performance in Complex Systems. In Proceedings of the Winter Simulation Conference IEEE, Phoenix, AZ, U.S.A., December 5th -8th ,1999.: vol.1, 815-820.

[3.7] VA: Virtual Design 2 - Programmers Guide 4.5.1. vrcom GmbH, 2006.

[3.8] PRESSER L., WHITE J.: Linkers and loaders. ACM Computers Surveys 4, 3 (Sept. 1972), 150–151.

[3.9] DREPPER U.: How To Write Shared Libraries. Red Hat, Inc., Aug. 2006.
http://people.redhat.com/drepper/dsohowto.pdf.

## CHAPTER 4

MANUAL WORK CELL SIMULATION IN A VIRTUAL ENVIRONMENT

ERGONOMIC OPTIMIZATION OF A MANUFACTURING SYSTEM WORK CELL IN A VIRTUAL ENVIRONMENT

The work deals with the development of a methodology for studying, in a virtual environment, the ergonomics of a work cell in an automotive manufacturing system. The methodology is based on the use of digital human models and virtual reality techniques in order to simulate, in a virtual environment, human performances during the execution of assembly operations. The objective is to define the optimum combination of those geometry features that influence human postures during assembly operation in a work cell. In the demanding global marketplace, ensuring that human factors are comprehensively addressed is becoming an increasingly important aspect of design. Manufacturers have to design work cells that conform to all relevant Health and Safety standards. The proposed methodology can assist the designer to evaluate the performance of workers in a workplace before it has been realized. The paper presents an analysis of a case study proposed by COMAU, a global supplier of industrial automation systems for the automotive manufacturing sector and a global provider of full maintenance services. The study and all the virtual simulations have been carried out in the Virtual Reality Laboratory of the Competence Regional Centre for the qualification of transportation systems (CRdC "Trasporti" - www.centrodicompetenzatrasporti.unina.it), which was founded by the Campania region with the aim of delivering advanced services and introducing new technologies into local companies operating in the field of transport.

4.1 INTRODUCTION

The implementation of a so-called "Digital Factory" is a tremendous challenge for automotive engineering. The technical task is to effect a seamless information backbone spanning three key departments: Design, Production Process Planning, and Manufacturing. Also suppliers such as machine and tool vendors have to be integrated into the information flow. Furthermore, there is the challenge of assimilating the human factor into the digital factory. New production planning tools will significantly change not only the contemporary production process planner's work but also the collaboration with suppliers. This raises one major issue: how to integrate different user groups into the design of complex engineering applications for production planning. The authors focus on a case study about the development of a methodology for optimizing the workplace in the automotive field. In particular they investigate the feasibility of integrating virtual humans into design environments to perform ergonomic assessments [4.1]. The paper illustrates the general benefits of ergonomic assessments, detailed advantages due to the utilization of virtual humans. A virtual human is an accurate biomechanical model of a human being. These models fully mimic human motion to allow an ergonomics (or human-factors) expert to perform process flow simulations. This study uses an analysis of the JACK software package to highlight the usefulness of such software options for applications in the manufacturing industry [4.2]. Workplace ergonomic considerations have traditionally been reactive, time-consuming, incomplete, sporadic, and difficult. The experience of an expert in ergonomic studies or data from injuries that have been previously observed and reported have always been necessary for these studies, and analyses are made after problems have occurred in the workplace. There are now emerging technologies supporting simulation-based engineering, and several operational simulation-based engineering systems to address this in a proactive manner. At present various commercial systems are available for ergonomic analysis of human posture and workplace design.

4.2   RELATED WORK

The importance of applying ergonomics to workplace design is illustrated by the Injuries, Illnesses, and Fatalities (IIF) program of the U.S. Department of Labor, Bureau of Labor Statistics [4.3]. According to this report, there were 5.2 million occupational injuries and illnesses among U.S. workers and approximately 5.7 of every 100 workers experienced a job-related injury or illness. Workplace-related injuries and illnesses increase workers' compensation and retraining costs, absenteeism, and faulty products. Many research studies have shown the positive effects of applying ergonomic principles in workplace design [4.4]. Riley et al. 4.5] describe a study to demonstrate how applying appropriate ergonomic principles during design can reduce many life cycle costs. Traditional methods for ergonomic analysis were based on statistical data obtained from previous studies or equations based on such studies. An ergonomics expert was required to interpret the situation, analyze and compare with existing data, and suggest solutions. The standard analytical tools included the NIOSH lifting equation [4.6], Ovaka posture analysis [4.7], and Rapid Upper Limb Assessment [4.8], among others. Various commercial software systems are now available for ergonomic studies. Hanson [4.9] presents a survey of three such tools, ANNIE-Ergoman, JACK, and RAMSIS, used for human simulation and ergonomic evaluation of car interiors. The tools are compared and the comparison shows that all three tools have excellent potential in evaluating car interiors ergonomically in the early design phase. JACK [4.10], an ergonomics and human factors product, enables users to position bio-mechanically accurate digital humans of various sizes in virtual environments, assign them tasks and analyze their performance. Gill et al. [4.11] provide an analysis of the JACK software to highlight its usefulness for applications in the manufacturing industry. Eynard et al. [4.12], describe a methodology using Jack to generate and apply body typologies from anthropometric data of the Italian population and compare the results with a global manikin. The study identified the importance of using accurate anthropometric data for ergonomic analysis. Sundin et al. [4.13], present a case study to highlight benefits of the use of JACK analysis in the design phase of a new Volvo bus. The importance of virtual humans in simulation and design has also been put set out Badler [4.1] & Hou [4.14]. Ford has made use of the "Design for Ergonomics" virtual manufacturing process, using JACK. The Ergonomic Design Technology Lab at Pohang Institute of Science and Technology is also involved in human modeling, design simulation, design evaluation in virtual environments and design optimization [4.15]. The potential value of ergonomics analysis using virtual environments is discussed in detail by Wilson [4.16].

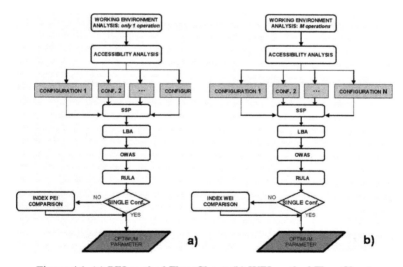

Figure 4.1. (a) PEI method Flow Chart; (b) WEI method Flow Chart.

4.3   THE METHODOLOGY FOR OPTIMIZING A WORKPLACE: PEI METHOD & WEI METHOD

In this work the problem that we have faced is optimization of the geometric features of a workplace in order to guarantee the maximum postural comfort for operators from different anthropometric percentiles during assembly operations. Such optimization, which has to consider the presence of possible external restraints, is strictly connected to the layout of the physical elements present in the working area. For this purpose, a methodology is proposed, based on the application of the "Task Analysis Toolkit (TAT)" included in the JACK software, whose functions will be analyzed in the following sections. Among the tools made available by TAT for the analysis of a working activity (NIOSH Lifting Analysis; RULA; Manual Material Handling Limits; Static Strength Prediction; OWAS; Low Back Analysis; Predetermined Time Standard) it has not been possible to find "one" that enables us to determine, among several solutions, the optimal one. If the geometric features characterizing a workplace influence the ergonomics of only one operation, in order to define the optimum combination of these features, the PEI method can be applied. It follows the phases illustrated in the flow diagrams of Figure 4.1a. The aim of the PEI method is the ergonomic optimization of an operation within a work cell, so it is referred to a single operation. In general, more than one operation is performed in a work cell. In this case, the combination of geometric features could not influence the single operations in the same way and, therefore, the PEI method is not applicable. Therefore, the combination of geometric features that optimizes the posture of all human percentiles can be evaluated applying the WEI method. Figure 4.1b shows the flow diagram of this approach, where M represents the number of operations that have to be performed in the work cell.

4.3.1   First Phase: ANALYSIS OF THE WORKING ENVIRONMENT

The first phase consists in an analysis of the working environment and in the consideration of all the possible movement alternatives: this, in general, involves considering alternative routes, postures and speeds of execution, which all contribute to the effective conclusion of the work. It is essential, in a virtual environment, to simulate all these operations in order to verify in the first place their feasibility. In fact, for instance, it cannot be taken for granted that all the points can be reached starting from different postures. The execution of this analysis guarantees the feasibility of the assignment. Among the phases of optimization this is the one that requires the longest time, since it needs the creation of a large number of simulations in real time, without taking into account that some of them will turn out to be useless, because, for instance, the simulation shows that some points cannot be reached with the movements that the designer had conceived. Other

64

parameters that can be modified are the distances of the manikin from objects taken as a reference, and the possibility to move the objects in the working area.

4.3.2 Second Phase: REACHABILITY AND ACCESSIBILITY ANALYSIS

The design of a workplace always requires a preliminary study of the accessibility of the critical points. This is a very interesting problem, and often occurs in assembly lines. The problem consists in verifying that in the designed layouts it is possible to carry out the movements necessary to the operation and that all the critical points can be reached; in a lifting operation, for instance, it could happen that a shelf is positioned too high and that therefore the worker does not succeed in developing his assignment. Such an analysis can be conducted in JACK, activating the collision detection algorithm. The layout configurations that do not satisfy the accessibility analysis do not have to be taken into consideration in the following analyses. From the analysis of the working environment and the accessibility analysis the different configurations can be designed. If the number of configuration is high a Design Of Experiments (DOE) procedure can occur [4.17].

4.3.3   Third Phase: STATIC STRENGTH PREDICTION (SSP)

Once the possible working sequences have been conceived, the question is: how many workers will be able to expound the necessary efforts for these movements? The answer can come from the Static Strength Prediction. In the case that the task must be developed, during a given period of time, by workers of different stature, age and sex, it can be accepted only in the hypothesis that the tool appraises in 100% the percentage of workers capable of the working activity. In practice, this cannot be done, because many activities provide percentages lower than 100%. In the workplace design phase, the operations that have a percentage of 0% should not be taken into consideration in the following analyses. The operations that have an evaluation of the percentage below a certain limit should also be discarded.

4.3.4   Fourth Phase: LOW BACK ANALYSIS (LBA)

Low Back Analysis is a tool that allows the strengths to be evaluated on the virtual manikin's spine, according to each posture assumed by the digital human model and any loading action. This tool evaluates, in real time, the actions linked to the tasks imposed on the manikin according to the NIOSH standards and according to the studies carried out in this field by Raschke [4.18]. The Low Back Analysis tool offers information related to the compression and cut strengths on the L4 and L5 lumbar disks, together with the reaction-moments in the axial, sagittal and lateral plane on the L4 and L5 lumbar disks and the activity level of the trunk muscles to balance the spine moments. In particular, in the following, we use the value, expressed in Newton, of the compression on the L4 and L5 lumbar disks.

4.3.5 Fifth Phase: OVAKO WORKING POSTURE ANALYSIS SYSTEM (OWAS)

OWAS is a simple method for verifying the degree of comfort related to working postures and for evaluating the degree of urgency that has to be assigning to corrective actions. The method was developed in the Finnish metallurgic industries in the 1970s. It is based on a classification of postures and on an observation of working tasks. The OWAS method consists in the use of a four-digit code to assess the position of the back side of the body, the arms and legs together with the intensity of existing loads during the performance of a specific task. The activity under examination has to be observed according a period of about thirty seconds. During each step, the positions and the applied strengths have to be registered, in accordance with a decomposing technique of complex activities. In this way, the distribution of the postures, the repeated positions and the critical positions are focused. The data collection and the successive analysis enable the working procedure to be redesigned to reduce or eliminate postures that are potentially dangerous. In fact, the tasks are classified using four principal classes: 1) no harmful effect, 2) a limited harmful effect, 3) recognized harmful effect on health, 4) highly harmful effect on health.

4.3.6 Sixth Phase: RAPID UPPER LIMB ASSESSMENT ANALYSIS (RULA)

From the initial scenario of possible layout configurations, the procedure progressively discarded those that: 1. did not ensure accessibility of the critical points; 2. asked for efforts that the workers were presumably not able to perform; 3. were potentially dangerous for the lower back. In this phase, the postural quality is analyzed. The purpose is to minimize the risks of muscular-skeletal pathologies in the medium-to-long term. The tool used is RULA. RULA analysis refers to exposure to risk of disease and/or damage to the upper limbs. The analysis takes into account loads, biomechanical and postural parameters focusing on the position of the head, body and upper limbs. The RULA method is based on data sheet filling. The sheet enables the user to quickly compute a value that indicates the degree of urgency of an intervention that needs to be adopted in order to reduce the risk of damage to the upper limbs. The method enables not only arm and wrist analyses, but also head, body and leg analyses. The first analysis, together with the information about muscles in use and existing loads, enables an assessment of the final score that represents the evaluation of the working posture. The risk is considered "acceptable" when the score is 1 or 2, "in need of further investigation" (a score of 3 or 4), "in need of further investigation and a rapid change" (a score of 5 or 6) or "investigation and immediate change" (a score of 7).

4.3.7   Seventh Phase: PEI EVALUATION

At this point a comparison can be established among the layout configurations, through the critical postures associated with them. The comparison allows us to establish a classification of risk of the operator contracting muscle-skeletal pathologies in the medium-to-long term. The choice of this optimal solution passes through the individuation of the more comfortable posture, which can be carried out using a Posture Evaluation Index (PEI), which integrates the results of LBA, OWAS and RULA [4.19]. In particular, PEI is the sum of three adimensional variables $I1$, $I2$ and $I3$. The variable $I1$ is evaluated normalizing the LBA value with the NIOSH limit for the compression strength (3400 N). Variables $I2$ and $I3$ are respectively equal to the OWAS index normalized with its critical value ("3") and the RULA index normalized with its critical value "5".

$$PEI = I1 + I2 + I3 \qquad (1)$$

where: $I1 = LBA/3400$ N, $I2 = OWAS/3$, $I3 = RULA/5$.

PEI definition and the consequent use of LBA, OWAS and RULA task analysis tools depend on the following consideration. The principal risk factors for work requiring biomechanical overload are: repetition, frequency, posture, effort, recovery time. The factors that mainly influence the execution of an assembly task are extreme postures, in particular of the upper limbs, and high efforts. Consequently, attention has to be paid to the evaluation of compression strengths on the L4 and L5 lumbar disks ($I1$ determination), to the evaluation of the level of discomfort of the posture ($I2$ determination), and to the evaluation of the level of fatigue of the upper limbs ($I3$ determination). PEI enables us to select the modus operandi to perform the disassembly task in a simple way. In fact, the optimal posture associated to an elementary task is the critical posture with the minimum PEI value. The variables defining PEI depend on the discomfort level associated with the examined posture: the greater the discomfort, the higher are $I1$, $I2$ and $I3$ and, consequently, PEI.

PEI expresses, in a synthetic way, the "quality" of a posture with values varying between a minimum value of 0,47 (no loads applied to the hands, values of joints angles within the acceptability range) and a maximum value depending on the $I1$ index. In order to ensure the conformity of the work with the laws protecting health and safety, a posture whose $I1$ index is more than or equal to 1 is assumed not valid. In fact, in this way the NIOSH limit related to compression strengths on the L4 and L5 lumbar disks will be exceeded. According to these considerations, the maximum acceptable value for PEI is 3 (compression strength on the L4 and

L5 lumbar disks equal to the NIOSH limit 3400 N; values of joints angles not acceptable). Iterating the procedure for all the elementary tasks of the assembly sequence, it is possible to associate to each of them the optimal posture to be assumed and, finally, to individuate the optimal value of the geometric parameters for the assembly task.

4.3.8   Eighth Phase: WEI EVALUATION

Once we have individuated the optimal value of the geometric parameters for each operation within a work cell (M represents the number of operation), the WEI (Work Cell Evaluation Index) index is introduced [4.20]. This is defined as:

$$WEI \left(Configuration_j\right) = \sum_i PEI_i * W_i \qquad (2)$$

where: = Time of Operation / Work Cell Time cycle.
The best index WEI is obtained by the following expression:

$$WEI_{BEST} = MIN_J \left[WEI(Config_j)\right] \qquad (3)$$

The WEI definition depends on the following consideration: if the aim is the ergonomic optimization of the work cell, it is necessary to establish a single optimal solution.

4.4   CASE STUDY

In order to test the PEI method and the WEI method, a case study proposed by COMAU was analyzed. The goal was to optimize a body welding work cell by using the methodology explained above.

4.4.1   Working environment analysis

The geometric model of the body welding work cell was imported into the JACK software, and then the 9 operations that have to be realized in the work cell were simulated. The 9 operations are as follows:

1.	Welding Visual Control	4.	Smearing Sealer with Gun	7.	Upper Cross member Wind Screen Loading
2.	Welding Imperfections Restoring with Brush.	5.	Smearing Sealer Renewal	8.	Back Cross member Wind Screen Loading
3.	Braze Welding Renewal	6.	Bottom Cross Member Wind Screen Loading	9.	Upper Front Rafter Loading

Figure 4.2 shows the sequence of operations performed in the body welding work cell. Simulating the operations that have to be performed in this work cell, a qualitative analysis of the postural sequence for each operation was conducted, in order to individuate the geometric parameters to be optimized. Table 4.1 shows the results of this first phase. As shown in the same table, a score was assigned to each operation and a calculation was made of the total score associated to each geometric parameter that influences the postural positions of the workers while they are performing the tasks.

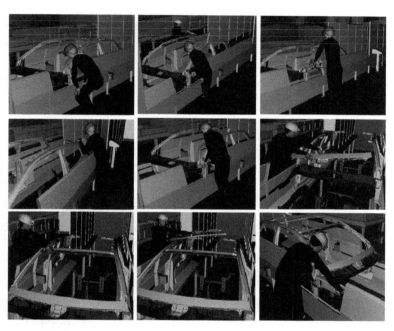

Figure 4.2. Sequence of operations in the Body Welding Work Cell.

Table 1: Definition of the parameters to be optimized

OPERATIONS / PARAMETERS	OP 1	OP 2	OP 3	OP 4	OP 5	OP 6	OP 7	OP 8	OP 9	SCORE
	1	3	2	1	2	3	2	2	4	$\sum_{i=1}^{9} s$
WORKER										
PERCENTILE	✓	✓	✓	✓	✓	✓	✓	✓	✓	20
POSTURAL POSITIONS	✓	✓		✓					✓	5
GEOMETRIC PARAMETERS BODY										
LOCKING POINT BODY	✓	✓	✓	✓				✓		7
HEIGHT BODY RESPECT TO ASSEMBLY LINE	✓	✓	✓	✓	✓	✓	✓	✓	✓	20

SCORE VALUE	MEANING
0	NO CRITICAL
1	LOW INJURY
2	MIDDLE-LOW INJURY
3	MIDDLE-HIGHT INJURY
4	HIGHT INJURY

It can be asserted that in this case study there is just one geometric parameter to be optimized, represented by the body height with respect to the assembly line, and one external factor, represented by the percentile of the worker.

4.4.2 Accessibility analysis

The analysis of the geometry was conducted in order to define the range of the body height, taking into account the geometric constraints of the work cell. Then, the range was reduced through the accessibility analysis.

The visual control of the welding (operation 1) and the smearing sealer with a gun (operation 4) defined the limits of the range, as shown in Figure 4.3 and Figure 4.4. The lower limit is -5cm, because the body positioned at -10cm does not allow the spot welding to be visualized completely (Figure 4.3b), and the upper limit is 20cm, because the body positioned at 25cm does not allow the smearing sealer to be realized by the 5th percentile. A step of 5cm was established, so the possible body height values are 6 (L1=6), and the percentiles considered are 3 (L2=3): 5th, 50th, 95th.

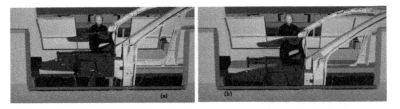

Figure 4.3. (a) Spot Welding visible; (b) Spot Welding not completely visible

Figure 4.4. The Smearing Sealer operation realized by the 5th percentile (on the left), vs the Smearing Sealer operation not practicable by the 5th percentile (on the right).

Now it is possible to define the number of configurations (N): from the combinations of the values of these parameters there are 18 configurations, as shown in Table 4.2.

70

Table 4.2. Experimental plane

CONFIGURATION N= L1*L2	HEIGHT BODY L1= 6	PERCENTILE L2= 3
1	-5	5°
2	-5	50°
3	-5	95°
4	0	5°
5	0	50°
6	0	95°
7	5	5°
8	5	50°
9	5	95°
10	10	5°
11	10	50°
12	10	95°
13	15	5°
14	15	50°
15	15	95°
16	20	5°
17	20	50°
18	20	95°

4.4.3 PEI Method & WEI Method Results

By applying the SSP, LBA, OWAS and RULA tools for each configuration and operation, the configurations injurious for the worker have been discarded. Table 4.3 shows results for the WEI METHOD & PEI METHOD. Note that PEI has been evaluated taking into account an average value among those obtainable. It can be asserted that the height of the body with respect to the assembly line, corresponding to the optimal postural sequence, is 20 cm (see Appendix, related to optimum solution reports). Table 3 shows the evaluation of PEI and WEI in some exhaustive cases.

Table 4.3. WEI METHOD & PEI METHOD results

HEIGHT BODY (mm)	Op1 W=.044 PEI	Op 2 W=.088 PEI	Op 3 W=.088 PEI	Op 4 W=.133 PEI	Op 5 W=.133 PEI	Op 6 W=.159 PEI	Op 7 W=.133 PEI	Op 8 W=.133 PEI	Op 9 W=.088 PEI	WEI
100	1.937	1.600	1.611	1.801	1.600	2.599	1.627	1.644	2.218	1.84
150	1.854	1.432	1.527	1.679	1.433	2.590	1.869	1.814	1.859	1.82
200	1.753	1.263	1.534	1.543	1.263	2.527	1.863	2.057	1.721	1.77

4.5   CONCLUSIONS

The proposed methodology makes available a valid tool for workplace analysis. The following objectives have been achieved: to appraise the quality of the postures assumed during a working activity; in designing a new layout, to establish if it ensures the feasibility of the operation (based on the criteria of accessibility of the critical points, of compatibility of the efforts, and danger for the lower back). to compare the possible alternatives for the configuration of the layout, supplying useful criteria for the designer to choose which is the most convenient to realize in the production chain. The reliability of the results depends on the extent to which the assumptions on which the tools of the TAT are based will be respected: almost static movements, non excessive temperature and humidity of the environment, satisfactory times of rest. Such assumptions are generally satisfied in the normal workplace. The objective of industry is to apply ergonomic criteria to reduce the number of accidents in the workplace and, secondly, to increase productivity. Currently only large firms turn their attention to this sector, because the simulation software has a certain cost, and also because the time for applying the software requires human resources that small firms do not have. The future objective is, on the one hand, to improve the interaction between the theoretical concepts of the ergonomics and the software, and, on the other, to simplify the analytical procedures to reduce time and costs.

ERGONOMIC EVALUATION OF MANUAL WORK CELLS TROUGH A DIRECT MANUAL INTERACTION APPROACH IN VR

In order to achieve an ergonomic evaluation of manual work cells through a direct manual interaction approach in VR it is necessary to integrate in a single visualization and development environment, tracking system (ART DTrack), manipulation systems (5DT Data Glove and CyberGlove), 3D navigation devices (SpaceBall, Flystick and Joystick) and stereoscopic visualization systems (Powerwall and Helmet such as Head Mounted Display). In this environment the user, who wears HMD, glove and tracking sensors, have the opportunity to execute, directly on the virtual model of the work cell, "subjective" ergonomic analyses simulating assembly, welding (Figure 4.5) and handling operations. The evident advantage of the direct manual interaction approach consists in the immediate correspondence between the desired analysis and the relative action carried out. This correspondence makes the performed analysis natural, intuitive, and, consequently , quick. Regarding the objectivity of the analysis results, the virtual manikin approach gives more flexibility because it easily allows several human models, corresponding to different percentiles, to repeat the same actions. Instead, the direct manual interaction approach presents, from this point of view, a lower correspondence to reality: in this case, in order to obtain more objective results, it is necessary to carry out a statistical analysis, through subjective tests on a representative sample of human population.

However, the approach with virtual manikins in the analyses of assembly or maintenance operations shows the disadvantage of moving the various degrees of freedom of the digital human model (135 for Jack of the EAI-UGS, 86 for Delmia - Ergo) by inadequate input devices like mouse and keyboard.

In Figure 8 an accessibility evaluation of a welding pincers is simulated by direct manual interaction through a virtual hand. By this method the evaluation is subjective as aforementioned, but it is possible to evaluate the layout configuration of the tools (the welding pincers) in order to improve the tasks of the operator.

Figure 4.5. Body Welding Work cell (on the left). Grasping of a welding pincers in a virtual environment (on the right).

The operation performed in this manual work cell consists in a manual braze welding renewal (Figure 4.5) of the spot welding realized in a previous automatic work cell by anthropomorphic robots. The position of a body welding pincer has been evaluated through the manual direct interaction in order to guarantee the maximum postural benefit of the worker. The creation of the virtual environment has been realized importing the geometries, the tools, and the equipment in VD2 using the VRLM format. The VR simulation has allowed us to evaluate in real time several layout configurations of the equipment in order to realize the grasping in ergonomic way and to define the best solution.

73

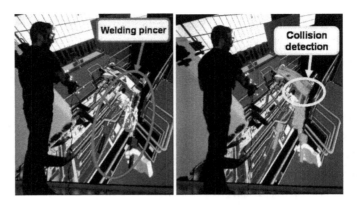

Figure 4.6. Collision detection during the welding operation.

4.6   REFERENCES

[4.1] Badler, N.: Virtual humans for animation, ergonomics, and simulation. In Proceedings of the IEEE Non Rigid and Articulated Motion Workshop, 1997, p. 28-36.

[4.2] Choi, H.: Integration of Multiple Human Models into a Virtual Reality Environment. Thesis, Washington State University. 2003.

[4.3] U.S. Department of Labor (Bureau of Labor Statistics) Workplace Injuries and Illnesses in 2001. Available online via <http://www.bls.gov/iif/home.htm.>[accessed January 2004].

[4.4] Das, B., and A. Shikdar. 1999. Participative versus assigned production standard setting in a repetitive industrial task: a strategy for improving worker productivity. International Journal of Occupational Safety and Ergonomics, 5(3): 417-430.

[4.5] Riley, M. W., Dhuyvetter, R. L.: Design cost savings and ergonomics. In Proceedings of the XIVth Triennial Congress of the International Ergonomics Association and 44th Annual Meeting of the Human Factors and Ergonomics Association, 'Ergonomics for the New Miennium', July 29-Aug 4, 2000, San Diego, CA USA.

[4.6] Dempsey, P. G.. Usability of the revised NIOSH lifting equation. Ergonomics, 45(12), 2002, p. 817-828.

[4.7] Keyserling, W. M. 2004. OWAS: An Observational Approach to Posture Analysis. Available online via <http://ioe.engin.umich.edu/ioe567/OWAS.pdf> [accessed January 2004].

[4.8] McAtamney, L. and E. N. Corlett. RULA: A survey method for the investigation of work-related upper limb disorders. Applied Ergonomics, 24(2), 1993, p. 91-99.

[4.9] Hanson, L.: Computerized tools for human simulation and ergonomic evaluation of car interiors. In Proceedings of the XIVth Triennial Congress of the International Ergonomics Association and 44th Annual Meeting of the Human Factors and Ergonomics Association, 'Ergonomics for the New Millennium', July 29-Aug 4, 2000, San Diego, CA USA.

[4.10] Di Gironimo G., Martorelli M., Monacelli G., Vaudo G.: Use of virtual mock-up for ergonomic design. Proc. of 7th International Conference on "The role of experimentation in the automotive product development process" – ATA 2001 [CD Rom], Florence, May 23-24 2001.

[4.11] Gill, S.A., R.A. Ruddle.: Using virtual humans to solve real ergonomic design problems. In Proceedings of the 1998 International Conference on Simulation, IEE Conference Publication, 457: 223-229.

[4.12] Eynard, E., et al: Generation of virtual man models representative of different body proportions and application to ergonomic design of vehicles. In Proceedings of the XIVth Triennial Congress of the International Ergonomics Association and 44th Annual Meeting of the Human Factors and Ergonomics Association, Ergonomics for the New Millennium', July 29-Aug 4, 2000, San Diego, CA USA.

[4.13] Sundin, A., M. Christmansson and R. Ortengren. Methodological differences using a computer manikin in two case studies: Bus and space module design. In Proceedings of the XIVth Triennial Congress of the International Ergonomics Association and 44th Annual Meeting of the Human Factors and Ergonomics Association, 'Ergonomics for the New Millennium', July 29-Aug 4, 2000, San Diego, CA USA. 496-498.

[4.14] Hou, H., S. Sun and Y. Pan.: Research on virtual human in ergonomics simulation. Chinese Journal of Mechanical Engineering, 2000, 13: 112-117.

[4.15] Ergosolutions Magazine 2003. Available on-line via http://www.ergosolutionsmag.com/ .

[4.16] Wilson, J. R. 1999. Virtual environments applications and applied ergonomics. Applied Ergonomics, 30(1) 3-9.

[4.17] Park S. H.: Robust Design and Analysis for Quality Engineering. London, Chapman and Hall, 1996

[4.18] Raschke U.: Lumbar Muscle Activity Prediction During Dynamic Sagittal Plane Lifting Conditions: Physiological and Biomechanical Modeling Considerations. Ph.D. Dissertation Bioengineering, University of Michigan USA, 1994.

[4.19] Di Gironimo G., Monacelli G., Patalano S.: A Design Methodology for Maintainability of Automotive Components in Virtual Environment. Proc. of International Design Conference - Design 2004, Dubrovnik, May 18 - 21, 2004.

[4.20] Love, R.F., Morris, J.K., Wesolowsky, G.O., Multiobjective & Methods, North-Holland, 1989.

## CHAPTER 5

AUTOMATIC ASSEMBLY LINE SIMULATION IN A VIRTUAL ENVIRONMENT

DESIGN OF AN INNOVATIVE ASSEMBLY PROCESS OF A MODULAR TRAIN IN A VIRTUAL ENVIRONMENT

In this work we present an innovative assembly cycle of railway vehicles that can improve the manufacturing process. The study was carry out using virtual reality (VR) technologies in the VR Laboratory of the Competence Regional Centre for the qualification of transportation systems set up by Campania Regional Authority. Using the developed simulation environment we were able to evaluate different workplaces layout configurations of the train assembly cycle. The best workplace layout configuration was detected in order to minimize the lead time in the production line and optimize the automation level and human component for each workplace.

5.1   INTRODUCTION

The work deals with an application that concerns the use of the simulation methodologies within the manufacturing process in the railway field thanks to the virtual reality (VR) technologies. The simulation, in fact, concerns a hypothesis of a "digital factory" developed to represent an innovative assembly cycle with regard to the criteria generally represent the state of the art in the material roadway manufacturing. The simulation offers the possibility to visualize the virtual assembly of a railway coach in 12 workplaces and a span of the shed in which the assembly is carried out.

Furthermore, it highlights the critical points on which the designers have to focus their attention in order to define an innovative solution able to ensure clear advantages in terms of feasibility and saving time and costs. The innovative manufacturing process introduces two types of problems. The first one is related to the necessity to reproduce the spaces and the equipment required to realize the manufacturing process. The second problem is to make clearly visible, through a realistic simulation, the handling of the parts inside the shed and the assembly sequences of the components and subcomponents, evaluating every geometric, functional and technological constraint.

5.2   STATE-OF -ART IN THE RAILWAY VEHICLE PRODUCTION

With a view to bringing innovations in the production cycle of a railway coach, it is necessary to examine with attention the different moments that lead to the design of a train, taking into account the techniques and technologies in use. A railway vehicle is essentially composed by two principal systems: the case and the bogies [5.1]. The case consists of a metallic structure of steel or aluminium alloy (in which four main modules are individualized: headstocks, body, body sides, imperial), the electrical equipment (illumination, signalling, power), the pneumatic system (brake and other services), the furnishing (seats, baggage racks, doors, dividing walls, lining), the auxiliary equipments (conditioning, hygienic services, fire-fighting) and, finally, the connection parts of the vehicles (hook-ups, repelling, electric and pneumatic couplers).

The bogie is, instead, composed by a metallic structure of steel (body), by the wheel arrangement, by the reduction gear (only for motive bogies), by the electrical equipment (signalling and control), by the pneumatic system (brake), by the parts of the suspensions (springs, dampers), by the auxiliary equipment (pneumatic sander) and, finally, by the interface parts with the case. The production flow of a railway vehicle is realized through four phases: component preparation; assembly structures; sandblasting and painting; final assembly (Figure5.1).

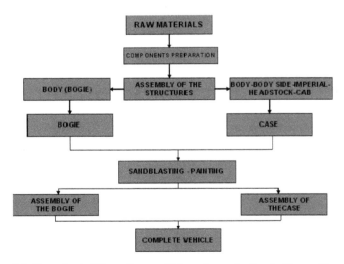

Figure 5.1. Flow chart of the railway vehicle cycle production in Firema Trasporti.

5.1.1 Component preparation

The raw materials are essentially constituted by sheets of various thickness (from 1 to 80mm), pipes of great thickness and diameter, circles full of great diameter, structural shapes and light sections of various forms and dimensions, some structural shapes have a length of over 30m. Beginning from such raw materials, the various components are drawn using traditional machines or to numerical control machines.

5.2.2 Assembly structures

After preparation, the components are assembled through welding or riveting in the department of carpentry. Having composed the different structures of the train skeleton (body, body sides, imperial and cab), the assembly of great dimensions is performed with the aid of a bridge crane. Here the various parts are welded and nailed to obtain the whole naked case. With an analogous process the bogies are obtained. The assembly finishes with the operations of calibration of the structures.

5.2.3 Sandblasting and painting

The body of the bogies and the steel cases are moved to the sandblasting department. The sandblasting plant allows the metallic surfaces to be prepared for subsequent operation of painting. Since the case in aluminium alloy does not need the sandblasting, it starts directly with operations of painting.

5.2.4 Final assembly

The case is fitted with systems and furnishing in a single workplace, with the tasks of two sets of skilled workers overlapping, and with materials differing in type and function such as electrical, pneumatic and mechanical equipments. These interferences are the main causes of the elevated production time of railway vehicles. The final phase of the railway vehicle manufacturing process consists of the lowering of the case on the bogies and of the carrying out of the final tests in order to satisfy the Specifications of Contract for the delivery to the customer. In this manufacturing process, about 80% of the time is needed for the assembly of the case. For this reason the railway industry is investigating the use of new technologies in order to improve the assembly process. This paper seeks to contribute to solving this problem and proposes to study innovative assembly cycles using simulation techniques in virtual environment

5.3   SIMULATION SOFTWARE SET-UP

In order to face the two categories of problems mentioned above, two software were used: the first one, Solid edge by UGS, was used for the three-dimensional modelling (Figures. 5.2, 5.3, 5.4), the second, Classic Jack by UGS [5.2–5.4] was used to reproduce, the handling to complete the various phases of the assembly cycle, verifying at the same time, the consequences of the possible interactions among the machines. It was thus possible to plan the assembly cycle so as to ensure the ergonomic requirements of the workplaces were met. With the software based on the use of digital human models, we were able to simulate the different work environments, the machines, the systems and, in the specific case, a whole factory, and to virtually reproduce operations of access, manipulation, assembly and disassembly, employment of tools and machines [5.5].

Figure 5.2. Digital model of a body module equipment named: "roasting-jack".

In recent studies conducted by our University research team in Naples in collaboration with experts in occupational medicine [5.6, 5.7], the advantages of the simulation, in virtual environment, of manual loads handling have been explained. In other works [5.8–5.10], the same research team have shown the digital human models features and the consequential advantages of new design methodologies development based on ergonomic simulation software. Use of such tools in simulating of worker's activities and of robotic systems makes it possible to perform ergonomic evaluations of the different workplaces when the project is still in its preliminary phase. An additional tool of Classic Jack, the Task Analysis Toolkit, is particularly useful for analysing working activities and comparing the results with the ergonomic limits imposed by the safety regulations.

Figure 5.3. Physical model and digital model of bridge crane.

Figure 5.4. Physical model and digital model of bracket crane.

The choice of the aforementioned software was dictated by the need to reduce to the minimum the data exchange problem between the CAD modelling environment and the virtual simulation environment. The selected software, both belonging to the family UGS, have the possibility of exchanging the data in optimal fashion [5.11], through the neutral format ".jt".

5.4 INNOVATIVE ASSEMBLY CYCLE OF A RAILWAY VEHICLE

The shed used to the new assembly cycle is constituted by 5 identical spans predisposed for the complete coach assembly. In each span there are 12 workplaces. Initially, the workstations layout proposed in the Figure 5.5 was considered valid. The study of the assembly sequence is developed in phase of concept, when it is possible to compare alternative solutions without creating limitations to the following phases. In the case study, since the shed is still being restructured, there is the possibility of operating any workplace layout configuration whatsoever. The same shed is divided in spans of 21m×170 m. Each of them is divided in 12 workplaces of equal area (7m×27 m). In order to transport the components toward the workstations or to transfer the completed modules (body, body sides, imperial, cabin and headstock) to the final assembly workplace, two independent bridge crane was adopted, with capability of lifting of 20 tons each. The independence of the lifting systems is useful in order to handle the components in the span.

Furthermore, two bridge cranes can be joined so that the capability of lifting is double. This solution results essential when the complete case is lowered on the bogies. In the Figure 5.6 is shown a top view of a shed span with the two bridge crane without the coverage. Each span is equipped with six bracket cranes; three of them are used by the first six workplaces, the other three by the second six. The bracket crane, applied to the wall, are mainly used for the handling of components to be assembled inside the workplace. In this case, the crane has a further degree of freedom: it can slide along the wall of the span so that it can be used by different workplaces, Figure 5.7. In the following sections each workplace of the shed, with the related tools to complete every assembly operations on the different modules, is described.

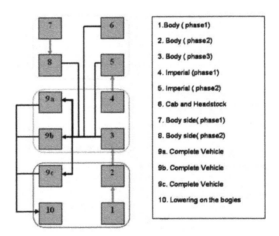

Figure 5.5. New configuration layout proposed for the railway vehicle manufacturing process.

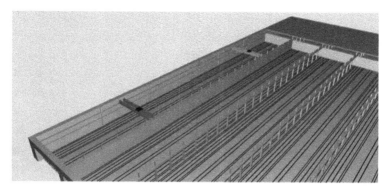

Figure 5.6. Shed top view.

5.4.1 Workplace 1

In workplace 1 the body module manufacturing begins, whose structure is realized in aluminium alloy 6005. The choice to use the aluminium alloy to replace the steel is well consolidated in the railway construction industry, not only to reduce the mass of the structure, but also to use extrusions of great dimensions, up to 25 m. Figure 5.8 shows the skeleton of the body as it reaches the first workplace.

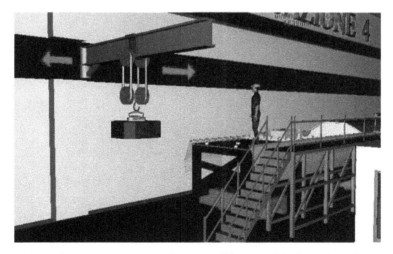

Figure 5.7. Bridge crane used for air-conditioner positioning on imperial.

The body must be positioned on an equipment built ad hoc, that accompanies the body module in every workplace. This module must be worked on both sides. In some assembly phases it is more convenient to work in standing position than in horizontal. The solution adopted allows work on both sides and in every possible configuration. Figures 9 and 10 show every characteristic mentioned above. Particularly, Figure 9 shows the body in vertical position with the central support closed. This support, in fact, is constituted by two separable elements.

The first workplace is also equipped with two anthropomorphic robots. For the particular operations to be developed, a specific type of robot has been individualized, the IRB 6600 Aseas Brown Boveri, whose characteristics are described in [12]. Such robots can be used in more than one workplace (as applied to the bracket crane), being assembled on bogies placed on rails integral with the floor.

In workplace n.1 are assembled in sequence every plant and equipment under the carriage case. Particularly, the positioning and the fixing of the electrical equipment and pneumatic system are realized.

These plants and equipments arrive packed and an operator positions them on the assembly jigs (the pipes are positioned, particularly, on special racks).

Figure 5.8. Body skeleton in aluminium alloy 6005.

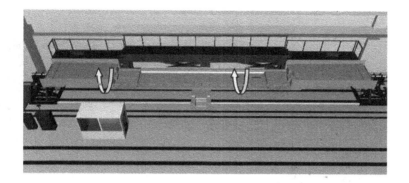

Figure 5.9. Body equipment: the red arrows show the translation direction of the equipment; the yellow arrows show the rotation around the axis of symmetry.

At this point the robots (with tools for the manipulation) withdraw the plants from the racks and position them on the body. In this phase is essential the support of a bracket crane for the correct positioning of the components on the body. In reality, for this operation, it can be considered the opportunity to use a portal robot to replace the bracket crane. Figures 11 and 12 show the two alternatives just described. However, the solution with the bracket crane results more flexible and versatile. On the one hand, in fact, a same crane can be used in more than one workplace, on the other hand, a portal has a greater overall dimensions and in much less agile in manipulating of components with considerable overall dimensions (as, for instance, the reservoirs of the pneumatic plant).

Figure 5.10. Body equipment rotated.

85

Figure 5.11. Portal robot solution. **Figure 5.12. Bracket crane solution.**

The virtual solution, represented in Figure 12, induces further considerations. There are, in fact, two operators that attend the pipes of the pneumatic plant collected in robots. Such a configuration results the most valid in the ergonomic terms. Indeed, standing operators, thanks to the presence of a mobile platform, can reach different heights without experience extreme fatigue in their arm muscles, when the body is in vertical position, Figure 13. Currently, in fact, the operators are forced to work in horizontal position, with consequent stress to the backbone.

When all the tasks in workplace n.1 are performed, the body module is complete of the electrical equipment and pneumatic system and of the whole componentry technology to them associated (pneumatic panel of auxiliary service, panel pneumatic brake, group production and treatment air, panel faucets back isolation, panel faucets anterior isolation, indicative of braked). The body includes other components (converter of traction, sand-blast reservoir, automatic coupler, bar of traction). Figure 14 shows the body module at the end-production of the first workplace.

Figure 5.13. Positioning of a pneumatic system pipe on the body .

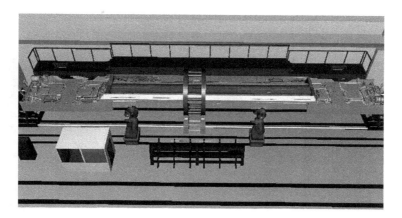

Figure 5.14. The body module at the first workplace end-production.

5.4.2 Workplace 2

In workplace 2 the body module is rotated of 180° to allow work on the opposite side of that worked in the previous workplace. This workplace is provided with two anthropomorphic robots ABB IRB 6600. The parts to be worked in this workplace are the three layers that constitute the flooring of the whole coach. Figure 15 shows the body complete with all the floor layers and, therefore, ready to be transferred to the workplace n.3. The flooring (Figure 16) is constituted by boarding in wood (1), by the rubber layers that covers the wood (2) and by the part in rubber that represents the trampling surface (3). Figure 17 shows in detail the body module at end-production of the workplace n.2.

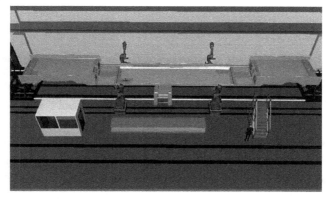

Figure 5.15. The body module ready to be transferred to workplace n.3.

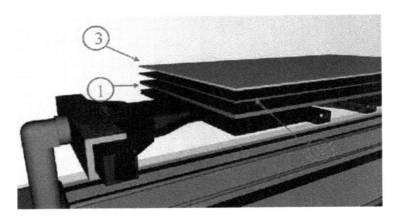

Figure 5.16. Flooring layers on the body.

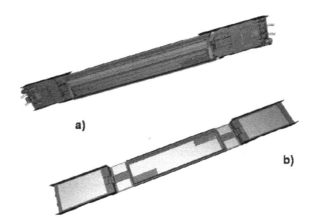

Figure 5.17. Body module completed: a) lower side; b) upper side.

5.4.3 Workplace 3

The body module, almost complete, is transferred to the workplace 3 where the seats and the handles are assembled. The equipment is positioned in such a way to allow the access to the paved body. Figure 18 shows the body module at the end-production of the workplace n.3. The seats assembly is completed using a ABB IRB 6600 robot. At this point, the body is transferred, through a bridge crane, to one of the workplace n.9 for the final assembly. Here it is left on a special structure where the operation of assembly take place with the body sides.

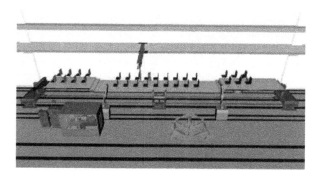

Figure 5.18. Body module at the end-production of workplace n.3.

5.4.4 Workplace 4

Workplace 4 is the first of the two workplaces dedicated to the assembly of the imperial module. The skeleton of the imperial is realized in composite material and it is the only part of the coach structure not realized in aluminium alloy. The technology of the composite materials is suitable for reproducing the imperial and, the costs of production are sufficiently reduced to maintain high the competitiveness of the end- product. The basic technique to be implemented in the whole imperial assembly has to ensure both the realization of a single frame, and more parts of smaller dimensions. Furthermore the specifications of the railway vehicle are currently regulated by the national and international standards (for instance, the EN12663) referring to the use of traditional materials (steel and the aluminium alloy). Use of the composite materials for the constructing the imperial is a solution that has to be tested in order to satisfy the standards and, also entails difficulty in realizing the connections between the imperial and the body sides. The Figure 19 shows the imperial module as it reaches, the workplace 4.

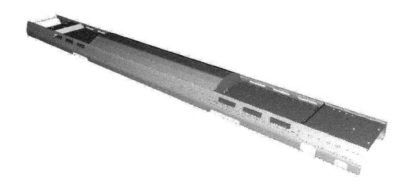

Figure 5.19. Imperial framework in composite material.

The equipment, that accompanies the imperial module along the workplace n.4 and n.5, is similar to that related to the body since the imperial also needs to be worked on both sides. We have to consider that, while the body module can be rotated easily by using roasting-jack, the imperial can not be rotated because of the weight of plants mounted upper the case such as the transformer that is the most heavy (1500kg). for this reason we need an assembly support built *ad hoc*. Figure 5.20a shows the assembly support built for the imperial module. The aim of this equipment is to guarantee the contact surface is as large as possible in order to avoid unwanted elastic deflection of the structure. Moreover, the assembly support is not set up directly on the floor but it is positioned on four sliding supports because the module, after the assembling of upper case plants, has to be translated to workplace n.5 with its support. Figure 5.20b shows the sliding supports just described and a mechanized lifting device that allows workers to assembly the equipment under case. In Figure 21 the following components have been assembled on it (component assembly upper case): air-conditioning cab (1), rheostat of brake (2), static converter (3), air-conditioning coach (4), inductor-condenser (5), asymmetrical pantograph (6) and hatch cover (7). For the assembly operations shown in the Fig. 21 can be used instead of the bracket crane a portal robot (Figure 22) that, in fact, it is particularly suitable for handling inside its workspace.

Figure 5.20a. Digital model of imperial assembly support.

Figure 5.20b. Sliding supports (on the left); mechanized lifting device (on the right).

5.4.5 Workplace 5

In workplace 5 the imperial module, coming from workspace 4, is set up with the components under the case. In this workplace the last operations are performed for completing the imperial packaging.

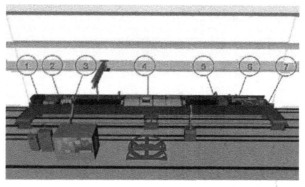

Figure 5.21. Assembly of all the components on the imperial.

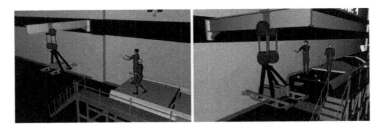

Figure 5.22. Workplace 4 with portal robot.

Figure 23 shows the imperial during the assembly of the air-conditioning canalizations and the electrical equipment (for the interior illumination of the coach).

Figure 24 shows the imperial ready to be assembled to the body sides. The imperial module, completed, is transferred to the workplace 9, through a bridge crane, where it is assembled to the body sides.

5.4.6 Workplace 6

The packaging of the cab module and of the headstock module is realized in workplace 6. The assembly of the hood, the furnishing wall and the door with the appropriate windows is performed in the headstock module.

Figure 5.23. Canalization assembly on the imperial.

The assembly of the following systems is performed in the cab module: pneumatic, electrical, air-conditioning, and furnishing. This makes it one of the most complex modules because of the high number of systems that it supports. The assembly steps needed to perform the complete preparation of the cab module in this workplace are illustrated in the following sections.

Figure 5.24. Imperial module completed.

Step 1

The first operation consists of the correct positioning of the resin shell on the building slip assembly. It is carried out with a bridge crane. The shell has to be positioned parallel to the platforms and aligned with the other parts of the case. In the following Figure 25 the resin shell and the relative support frame are respectively illustrated: Figure 26 shows the shell positioning on the building-slip assembly, while the employment of the platform truck for installing the cab components is shown in the following sections.

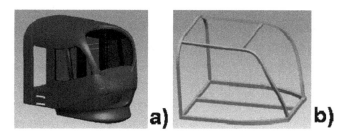

Figure 5.25. a) Resin shell; b) resin shell frame.

Figure 5.26. Shell positioning on the building-slip assembly.

Step 2

The following operation, immediately performed, is the covering of the walls with deadening panels. These panels are applied to all the surfaces with the purpose of reducing the sound frequencies. This intervention has a tendency to decrease the noise level in the cab, contributing to an improvement in comfort. The deadening panels can be composed of fibres or foams, with protective and aesthetical finishes that improve the acoustic characteristics. In such a way, the covering allows the shell to be isolated from all the vibrations and the noises of the train in motion.

Step 3

Having completed the assembly of the deadening panels, the following operation consists of the assembly of the monoblock system with the related ventilation channels and air conveyer. Normally, this operation is realized through the use of a bridge crane and requires that each component has to be assembled in the cab by an operator not working in ergonomic conditions. The innovation of this assembly step consists in dedicating a specific area to the assembly of the whole monoblock system, with the related ventilation channels and air conveyer. In this way, in workplace 6, the whole monoblock system can be positioned in the cab through a single operation. This operation, however, introduces some difficulties in the large-size part assembly within the shell. In order to insert, assemble and install the components inside the cab under the roof, a platform truck has been designed, to improve the postures of the workers. This truck allows the components to be centred automatically inside the shell.

The assembly and fixing of the ventilation channels can be transferred from the current manual labour to machine control. The human component can be reduced by introducing an anthropomorphic robot for the assembly and fixing of the components dedicated to the air conditioning system in the upper part of the cab. For the particular operations to be developed, a specific type of robot has been identified, Asea Brown Boveri by ABB. The work area is very wide since the robot can bend backwards completely. Typical areas of application are the spot welding, the manipulation and the components fixing [5.12].

This type of robot is used for the handling and screwing of the parts in this application. The air-conditioning assembly through the use of the new platform truck is illustrated in the following Figure 27.

Step 4

Having assembled the systems under the roof of the cab, the cables and the pipe clamps are assembled both on the floor and on the wall (Figure 28). The pipe clamp assembly can be automated through the use of the same robot as in the previous changing tool operation. The fireproof system is positioned in the tool cabinet in the cab and is structured for monitoring the presence of smoke inside the railway-truck [5.12].

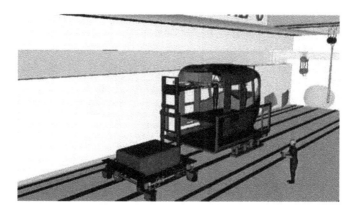

Figure 5.27. Positioning and assembling of the components under the roof of the cab.

Step 5

In this step, the assembly of the pneumatic system, the electrical equipment and the auxiliary equipment is realized manually. The main components are: pneumatic brake controls, suspensions controls, pantograph controls, flexible connections between case and bogie, filters, valves of retention, door handles, windshield wipers, illumination and fireproof cartridges.

Figure 5.28. Assembly and fixing of the cables.

Step 6

The following operation consists of the assembly of the control bench. The control bench, like the other systems, requires the fixing of all the components in a dedicated workplace. The whole control bench is then assembled and connected inside the shell in workplace 6. Before being assembled in its position, the bench is tested by special computers that verify the efficiency and the regularity of all the electric and pneumatic connections. The positioning of the control bench components (Figure 29) satisfies ergonomic requirements, such as visibility, reachability of controls and comfortable postures. These characteristics have been verified using the same virtual environment adopted for the assembly simulations, by means of virtual manikins, Figure 30. Since the cab module is still in the concept phase, it is possible to realize, in an entirely innovative way, two types of controls bench in the same workplace: with the right-hand drive and with the left-hand drive. Figure 31 shows the positioning of the bench through the platform truck.

Step 7

The following step consists of the floor assembly inside the cab. Figure 32 shows the components of the floor layers: 1 layers perforated to allow joining to the frame; 2 rubber carpet; 3 insulating panel; 4 cushioning supports; 5 multi layer wood; 6 cables.

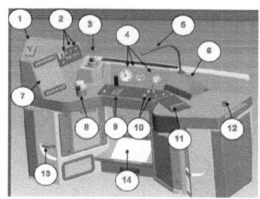

1) Calling Pushbutton Engine Driver
2) Switch Insulation Bench-up/Down Pantograph
Position-static Group -Extra Fast Switch
3) Modular Brake
4) Manometers
5) Microphone
6) Monitor SCMT
7) Monitor DIS
8) Electrical-pneumatic Brake
9) Direction Control
10) Pushbutton Of Trumpet And Whistle
11) Phone Earth Train
12) Help Push-button
13) Pneumatic Insulation Of Brake
14) Footrest Platform.

Figure 5.29. The control bench components.

Step 8

Having concluded the flooring in the cab, the seat and electric closet assembly is realized. The transport of the seats and the electric closets is performed through the employment of the platform truck. Once the definitive position is established, the seats are screwed by the workers and the electric cabinet is connected through the electrical connectors and then fixed to avoid possible movement. Figure 32 also shows the seat and electric cabinet assembly realized by the workers and the criteria of transport and anchorage of the electric cabin.

Figure 5.30. Visibility and reachability analysis using virtual manikins.

97

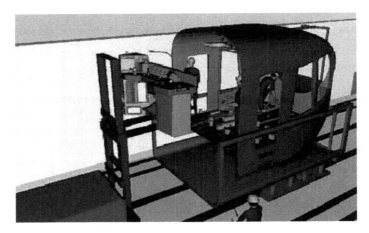

Figure 5.31. The control bench positioning inside the shell.

Figure 5.32. Floor layer assembly in the cab (on the left); transportation of the electric cabinet (on the right).

Step 9

The last operations to be realized to complete the assembly cycle of the cab module are the assemblies of doors, side windows, radiant panels, walls panels, frontal windscreen, headstock and door headstock. Figure 33 shows the use of a robot for the positioning of the side windows on the shell. The same Figure shows the completed cab module, ready to be connected to the coach. The connection between the cab module and the case involves mainly the electric and pneumatic systems. The air-conditioning system in the cab is generally independent from that of the passenger coaches.

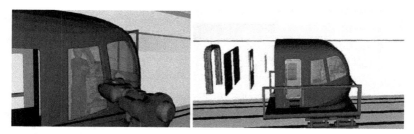

Figure 5.33. Assembly of the side windows through the robot ABB IRB 6600 (on the left); cab module completed.

5.4.7 Workplace 7

In workplace 7, and in the following, the assembly of the body sides is performed. Each of the two body sides of a coach has to pass through workplaces 7 and 8 in order to be equipped with every component. The structure of the body side is in aluminium alloy 6005. In order to make easier its motion, it is opportune to divide the body side in three parts, as shown in Figure 34.

Figure 5.34. Body side frame.

Figure 5.35. Workplace n.7.

The three parts of the body side are positioned on cribs with bending that reproduces its shape. Particularly, in workplace 7 (Figure 35) is realized the assembly of: deadening panels; windows; air-conditioning canalizations of the passenger coach. The first operation to be performed in the workplace 7 is the deadening panel assembly, that cover the extrusions of the body side. Currently the panels are positioned by the workers and fixed through small nails welded on the case that, once folded up, maintain the panels adherent to the structure. The number of these nails is considerable, so it is easy to imagine that this operation requires a consistent working time. For this reason, the assembly of the deadening panels has been automated and it consists of the following steps:

- panel collecting, (the panels are already shaped and are positioned behind two ABB IRB 6600 robots, on one of which a sucker tool is fixed) (Figure 5.36);
- panel positioning on the extrusions of the body side;
- nail closing on the panels and therefore fixing of the same nails on the extrusions, through another tool positioned on the robots.

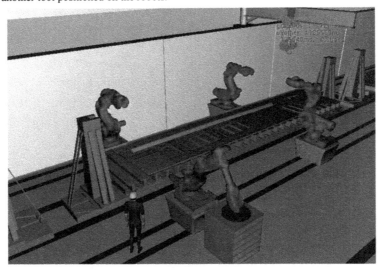

Figure 5.36. Panel positioning on the body-side.

Having finished the assembly of deadening panels, the second operation is the window fixing inside the coach. It is characterized therefore by the following main phases:

- window positioning by two workers on pallet prepared behind the ABB IRB 6600 robots;
- preliminary drilling of windows with the purpose to allow the screwing and the riveting:

100

- window collecting through the robots with the sucker tool and positioning in the window opening:
- screwing and riveting of the windows through the robots.

In this phase the sealing operation between the case and windows, will be performed after assembly sequence and particularly when the body side will be positioned on the body. Therefore the windows assembly is easily made automated due the repeatability and the high number of windows (six windows for the left broadside and eight for the right are assembled).

Subsequently, the other operation that is performed in workplace 7 is the fixing of the air conditioning canalizations. This phase consists of:

- collecting of the air conditioning canalizations through a bracket crane, handled by an operator through a pushbutton panel;
- lowering of the canalizations on the body side;
- collecting of the canalizations by the operators;

Figure 5.37. Aspiration of shaving and residues at the end-production.

- positioning of the canalizations on the body side with the purpose to mate the holes on the hanger of the canalizations and the holes situated on the structure; such operation allows the automation in the fixing process, facilitating the riveting;
- fixing of the canalizations on the structure through the rivet positioning on the hangers of the canalizations.

101

This operation is performed by the robots. Finally, before passing to the description of the workplace 8, a further operation of cleaning up is performed in order to remove shavings and residues, Figure 37.

5.4.8 Workplace 8

Workplace 8 is the second and the last workplace dedicated to the body side module. In this workplace are performed particular operations that request the use of two anthropomorphic robots (ABB IRB 6600) and of other two (ABB IRB 4400) [5.13] for silicone and scoring. The first phase is the positioning of the raceway cables.

To date, high and medium tension cables have to be manually positioned inside small canalizations, with long times. For this reason has been adopted the raceways in order to
guarantee the disassembly and the modularity and the maintainability.

Such operation is constituted by the following phases:

- positioning of the raceways on a rack: this operation is performed by two operators;
- collecting of the raceways: this operation is performed by two robots;
- positioning of the raceways on the hanger assembled on the structure, with the support rotated of 90°;
- fixing of the raceways through riveting;
- positioning of the cables inside the raceways (Figure 38).

The following phase is constituted by the assembly of the furnishing, by the lowering of the components on the body side through a bracket crane, by the collecting of the components, by the positioning of the components on the structure and by the relative fixing through riveting operation. The last operation is finally characterized by the assembly of the baggage rack, by its lowering on the furnishing (Figure 39), by the positioning and fixing through riveting. In the Figure 39 is shown the body side module completed.

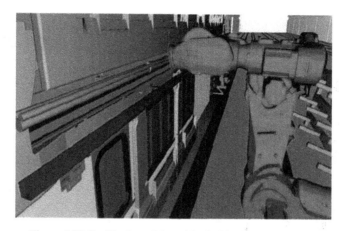

Figure 5.38. Positioning of the cables inside the raceways.

5.4.9 Workplaces 9a-9b-9c

At the beginning, the cycle production was conceived with an arrangement of three workplaces (number 9, 10, 11) for the final assembly, in which, first, the assembly between the body and the body sides was performed (workplace 9), then, the partial structure was transferred to the following workplace for the lowering of the imperial (workplace 10), and finally, the headstock and the cab assembly were completed (workplace 11).

This solution produced a useless duplication of resources in the workplaces used for the final assembly; for this reason three identical workplaces, renamed 9a, 9b and 9c, have been conceived. In this way, in each of these three workplaces, the body arrives from workplace 3, the body sides arrive from workplace 8, the imperial arrives from workplace 5, and the headstock and cab arrive from workplace 6. These three workplaces can be equipped on both sides with two anthropomorphic robots.

Figure 5.39. Lowering of the baggage rack (on the left); workplace 8.

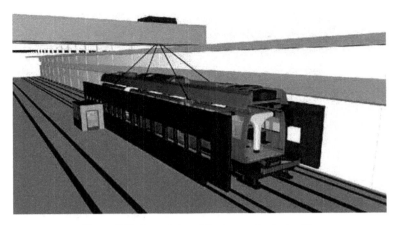

Figure 5.40. Imperial positioning on the body side.

Even if this proposed solution can reduce the number of vehicles produced in every span, the railway market does not have the problem of saturating the innovative production plant. The black arrows in Fig. 5 indicate that the workplaces 3, 5, 6 and 8, can in different ways converge into workplaces 9a, 9b or 9c.

The movements indicated by such arrows are performed with bridge cranes. In these workplaces the assembly of the body sides to the body is carried out through riveted and/or glued joints by using the anthropomorphic robot ABB IRB 6600. If the traditional welding technique was used to assemble the different parts of the case, there would be a risk, on the one hand, of having to restore the welding imperfections and on the other hand, of ruining parts of the furnishing or of other components situated on the structure of the module.

The selected solution provides for the positioning of the body sides with a bridge crane, on assembly jigs, where they are picked up by an ABB IRB 7600 robot [5.12], used for the manipulation of heavy components. At this point, two other ABB IRB 6600 robots perform the operations of drilling and riveting. The imperial is lowered onto the body sides with a bridge crane, Figure 40; the body sides are positioned at the correct distance thanks to the presence of adjustable interior supports. The extremities of these supports are in Teflon to avoid damage to the interior coach. Having been put in the correct position, the imperial is fixed to the body sides by riveting/glueing. The case assembly is concluded when the headstock and the cab arrive from workplace 8 by the use of a bridge crane. These two modules are fitted to the partial structure of the case, by glueing/riveting. Figure 41 shows the completed coach.

5.4.10 Workplace 10

In workplace 10, the complete coach, removed from the supports, is lowered onto the bogies by a bridge crane.

Figure 5.41. Railway coach.

5.5 CONCLUSIONS

This work mainly aims to offer a vision of the potentialities, still not completely fulfilled, that the simulation technologies based on VR can produce in order to improve the design. The costs to be sustained for using the VR in design are considerable as are plant and management costs due to the continuous demand for updating facilities and operators. For this reason the applications concretely realized to date are related to industrial compartments such as the automotive and aeronautical industry that are able to sustain the costs. Finally, the design methodologies based on the VR produce two other positive results. The first one is that they offer the designer freedom to invent different new solutions, since the tool that they have available allows them the contextual experimentation. The second advantage is that the design with the aid of the VR necessarily require the meticulous analysis either of the functional requisite, or of the project parameters, or of the constraints. These considerations, that spring not only from the necessity to model and to simulate the objects, but also from the simulation and the experimentation of their functionalities and their use, afford more detailed knowledge of the project. Such an advantage, would already represent an appreciable result.

DESIGN AND DEVELOPMENT OF A LIGHT AIRCRAFT ASSEMBLY CYCLE IN A VIRTUAL ENVIRONMENT

The work deals with the design of the assembly cycle of a light aircraft in a virtual environment. To date the methodologies of product development have remained traditional. In other words this does not give a competitive advantage to industries that are still on the market with aircraft similar in terms of class and typology. For a long time, the strategic objective of the industries of the aeronautical sector has been (and continues to be) to reconcile high quality and a high level of innovation with a low cost of development and manufacturing. Through the use of virtual reality it is possible to connect the phase of development of the aircraft with that of manufacturing. This study has been carried out in collaboration with OMA SUD Sky Technology using Virtual Reality (VR) technologies in the VR Laboratory, *VRTest*, of the Regional Competence Centre for the assessment of transportation systems founded by Campania Regional Authority. Thanks to VR techniques we have been able to identify the best workplace layout configuration for the subgroup production, to define the most efficient assembly sequence of the structural subgroups and to determine for one of the aircraft subgroups an assembly sequence that can optimize the assembly cycle and the operations performed by the workers in the first phases of design. All this has allowed us to resolve, in the phase of the design of the assembly line and the equipment for the assembly, a series of problems relating to the sequences of subgroup assembly and of the assembly line, concerning the final assembly of the aircraft structure. Through this methodology we have been able to test and to evaluate the interaction between human beings and machines before setting up the whole production line and creating the physical mock-up of the equipment. Therefore we have been able to propose design changes to the equipment before the beginning of the manufacturing process in order to make the operations to be performed as easy as possible and to reduce the time-cycle of the subgroups and the manufacturing costs.

5.6   INTRODUCTION

The work deals with an application that concerns the employment of simulation methodologies within the manufacturing process in the aeronautical field thanks to Virtual Reality (VR) technologies. The simulation concerns an hypothesis of the "digital factory", realized in order to represent an innovative assembly cycle in comparison to the criteria that generally represent the state of the art of aircraft manufacturing. The simulation, in fact, offers the possibility to visualize the virtual assembly of a light aircraft in ten workplaces and a span of the shed in which the assembly is realized. Furthermore, it allows us to emphasize the critical points on which the designers have to focus their attention in order to define an innovative solution able to assure clear advantages in terms of feasibility and time and costs saving [5.13].

The innovative manufacturing process introduces two types of problems. The first is related to the necessity of reproducing the spaces and the equipment required to realize the manufacturing process. The second is how to make clearly visible, through a realistic simulation, the handling of the parts inside the shed and the assembly sequences of the groups and subgroups, evaluating every geometric, functional and technological constraint.

5.7   LIGHT AIRCRAFT PRODUCTION CYCLE

In the following section the production cycle of a light aircraft is described, starting from the raw materials up to the finished aircraft. Figure 5.42 represents the production flow of an aircraft as it usually happens. An aircraft is essentially composed of structural systems and equipment. In the structure of an aeroplane we can distinguish three structural groups and within each group various subgroups. The three structural groups are, [5.14]: fuselage, wing and empennage. The fuselage group is composed of the following subgroups: the front frustum, central frustum, back frustum, doors and fairing. The wing group is composed of: the central caisson, leading edge, flap, ailerons and fairing. The empennage group is composed of: the vertical planes (the drifts and the rudders), horizontal planes (the stabilizer) and the trim. We can finally distinguish the front cart and the back or principal cart. Among the equipment we can identify: the motor fuel system, brake system, hydro plant, heating plant, air conditioning plant, vacuum and electric plant. We need to add to this list the motors, the furnishing, the dashboard and the bridge of command. In the following sections the assembly line of an aircraft is described in detail.

5.7.1   Component preparation

The raw materials essentially consist of aluminium sheets of various thicknesses, printed parts, steel outlines of various forms and dimensions. Beginning from such raw materials, through hydraulic presses and numerical control machines, the various components are designed and will constitute the subgroups of the aircraft (longerons, ribs and skin). In the following sections the different operations for obtaining the components are described: cutting, bending, grinding and drilling through the use of presses, moulding and surface treatments.

107

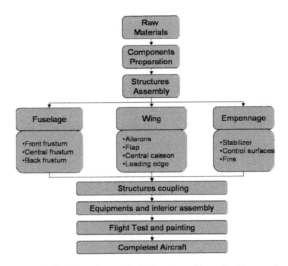

Figure 5.42. Flow chart of a light aircraft production cycle.

5.7.2   Subgroup assembly

The components coming from the preparation activities arrive at the store where the assembly kits of the different subgroups are prepared (fuselage frustum, caisson of the wing, ailerons, rudders and flap). By assembly kit we mean all the components that compose the single structural subgroup. The elements of the subgroups are assembled and joined together through riveting. In particular, each subgroup is assembled in a workplace along an assembly line, in which there is an assembly structure built ad hoc. We need to specify that the subgroups are assembled in parallel, taking into account the time needed, in order to avoid stoppages not considered in the assembly cycle that could cause delays in the production line. The flow time depends on the type of subgroup and on the complexity of the structure and therefore every workplace will have a different flow time.

5.7.3   Master group coupling of the aircraft

Once the different structures of the skeleton of the aircraft are completed, the frustum of the fuselage, wing and empennage, the final assembly is performed with the aid of a bridge crane, where the various parts are joined and nailed together in order to produce the complete casing of the aircraft. The last phase of the structure assembly is the assembly of the carts on the fuselage to allow the structure of the aircraft to be handled more easily between the following workplaces dedicated to the assembly of the equipment. Figure 5.43 illustrates the subgroups of a traditional light aircraft.

5.7.4 Equipment assembly

Once the structural assembly of the aircraft is completed, the motor and helix assembly is performed. Next the fuselage is fitted with systems and furnishing in various workplaces, with the tasks of different sets of skilled workers overlapping (i.e. pneumatic system fitters, air conditioning assembly technicians, window fitters and door assembly technicians)and with materials differing in type and function, such as electrical, pneumatic and mechanical equipment and furnishing panels. At this point the aircraft is completed and ready to be submitted for the flight tests, as prescribed in the standards relating to aircraft manufacturing.

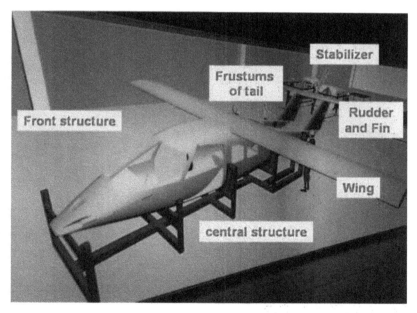

Figure 5.43. Subgroups of light aircraft.

5.7.5 Painting

Once all the tests on the finished aircraft are completed, the painting phase follows, in specific rooms fitted with special painting equipment.

5.8   SIMULATION SOFTWARE SET UP

In order to face the two types of problems previously mentioned, two software programs have been employed: the first, CATIA V5 by Dassault Systemes, has been used for the three-dimensional modelling, Figure 5.44; the second, Classic Jack by UGS, has been used to reproduce the handling for completing the different phases of the assembly cycle, verifying, at the same time, the consequences of the possible interactions between the machines. In such a way, it has been possible to plan the assembly cycle also guaranteeing the ergonomic requirements of the workplace. The software, based on the employment of digital human models, allows us to simulate the different working environments, machines, systems and, in this specific case, a whole factory, and to reproduce virtually the access, manipulation, assembly and disassembly operations and the employment of tools and machines, [5.15-5.19].

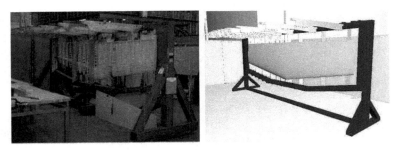

Figure 5.44. Physical model and digital model of stabilizer.

5.9   DESIGN OF A LIGHT AIRCRAFT ASSEMBLY CYCLE

The OMA Sud industry shed is composed of five spans and a central body with several floors in which the offices are situated. The production of the Sky Car will involve, at least in the initial phase, only one of these spans because the other four are used for other types of production.

In the following sections we describe the organization of the area dedicated to the production of the structural groups of the Sky Car, Figure 5.45. We define each workplace, the assembly sequence of the final structure and the assembly cycle of the subgroup rudder, considering that to date the equipment design has been completed only for the assembly of the vertical and horizontal planes of the tail.

The objective of the study is:

•   to define the workspace for the production of the subgroups;

110

- to define the assembly sequences of the structural subgroups;
- to determine for one of the subgroups of the aircraft an assembly sequence able to optimize the time-cycle and the operations performed by the workers from the first design phases of the equipment.

The assembly activities are simulated using a virtual reality environment. Virtual simulations allow us to resolve a whole series of problems relating to the assembly sequences of the subgroups and the assembly line, in particular relating to the final assembly of the aircraft structure. Furthermore, they allow us to test the assembly slips in the design phase, evaluating the interactions between human beings and equipment, and to propose tool changes in order to make the operations as efficient as possible. It is necessary to emphasize that the global assembly sequence is still in the concept phase and therefore, it is possible to propose alternative solutions without creating excessive "up settings" for the following phases. In particular, to date, the equipment used in other types of production has still been located in the span. For this reason we can propose possible hypotheses with the purpose of achieving the most convenient layout configuration of the workplaces from a logistic and economic point of view.

The span of the shed involved in the production of Sky Car has a length of 60 m and a width of 20 m.

In order to transport the components towards the workplaces in each span or to transfer the complete groups (front frustum, central frustum, back frustum, wing) to the workplace dedicated to the final assembly, bridge cranes are used, each with a capability of lifting of 50 tons. Each bridge crane is radio-controlled by an operator with a complete system of radio controls (a transmitter and a receiver).

Figure 5.45. Configuration layout proposed for Sky car assembly process.

111

5.10 DESIGN OF ASSEMBLY SLIPS FOR SUBGROUPS

In order to design and develop the new assembly process of the Sky car aircraft model, we have conceived the assembly slips for the aircraft subgroups in order to optimize the whole assembly cycle of the aircraft, satisfying ergonomic and functional requirements. For example, we describe the design phases carried out for the rudder assembly slip, (see figure 14). In the first phase, we carried out an analysis of the subgroup which allowed us to identify the sequence of the elementary operations that have to be executed in order to assemble the rudder according to functional criteria and geometrical constraints: a) assembly kit check, b) main spar positioning, c) saddle closing and fastening positioning, d) nose-rib positioning and blocking, e) rib positioning and blocking, f) sub spar positioning, g) interior structure drilling, h) positioning and riveting of hinges and supports, i) cross rib assembly, j) covering.

In the second phase, an accessibility and manipulability analysis was carried out, using the Jack software for ergonomic simulation by UGS. Each elementary operation was analyzed according to the defined sequence. For example, task (d), positioning and blocking of the nose-rib, is shown in Figure 5.46. In this case, the analysis highlighted that the access of the right hand and, consequently, the grasping of the element are performed without collision between a segment of the hand and the saddle. In the same figure the assembly kit is shown, positioned on the workbench. The choice of the optimal postural position was carried out using a Posture Evaluation Index (PEI) that integrates the results of the Low Back Analysis (LBA), Ovako Working Posture Analysis System (OWAS) and Rapid Upper Limb Assessment Analysis (RULA), [5.20-5.23].

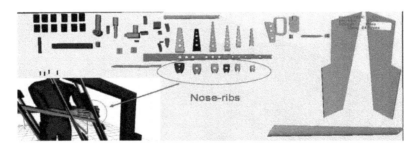

Figure 5.46. Assembly kit and nose-rib positioning.

5.11 WORKPLACES OF THE AIRCRAFT SUBGROUPS

In this phase it is necessary to define which workplaces will be dedicated to the structural assembly of the aircraft and the equipment to be placed inside. The assembly cycle consists of: assembly bench, assembly slip and assembly line. By assembly bench we mean those production activities that do not require particular equipment and that are carried out on normal work benches. The assembly slip includes those activities that are carried out with the aid of special equipment, built ad hoc for the type of subgroup to be assembled. All the activities of the assembly subgroups take place in dedicated workplaces. In the following sections we describe each workplace and the related equipment involved.

5.11.1 Workplace 1: front frustum

The assembly of the front frustum takes place in workplace 1. The assembly slip consists of two bodies: a base and a revolving body. The whole slip will have a height of 1.6 m, a length of 4.1 m and a width of 2.5 m. Since some components of the front frustum are pre-assembled at the bench and then positioned on the slip, two workbenches are present in this workplace. The tools to be used will have saddles of reference for the positioning of the axes of the frustum, of the lower beams of the layer and of the ribs.

Notice that the body consists of three main beams and is open in the upper part to allow, once the assembly is completed, an easier handling of the complete frustum with the bridge crane, Figure 5.47. In particular, the subgroup, once completed, is slung to the bridge crane with special ropes and transported to the workplace where the joining to the other subgroups will be carried out.

5.11.2 Workplace 2: central frustum

In workplace 2 the assembly operations of the central frustum of the fuselage are carried out, Figure 5.48. There are the assembly slip of the subgroup and a workbench. The assembly kit is brought to this workplace from the store and is checked and the presence of all the components that must be assembled is verified. Some components of the assembly kit are assembled on the bench and then positioned on the assembly slip where the central frustum is completed. The assembly slip consists of two bodies, a base and an upper body. After the subgroup assembly and verification are completed, at the end of all the control activities and testing, the central frustum is slung and transported to workplace 10, where the final assembly of the fuselage is carried out.

5.11.3 Workplaces 3 and 4: frustums of tail

Workplaces 3 and 4 are dedicated to the two frustums of the tail respectively. There are two assembly slips: one for the right frustum and one for the left. Both assembly slips are perfectly symmetrical and consist of two parts: the base and the upper body, Figure 5.49. On the upper side of the slip, five saddles and related supports are fixed and used as references for the positioning of the ribs. Two of the five saddles are fixed to the extremities of the slip; the other three are removable and are fixed to the body by screws and bolts, Figure 5.50. The saddles and the supports are removable to allow the operations of drilling and riveting in the lower part of the frustum. The upper body can rotate around its greater axis: this makes the assembly operations easier.

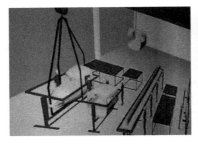

Figure 5. 47. Workplace of front frustum.

Figure 5.48. Workplace of central frustum.

5.11.4 Workplace 5: wing

The wing is the bulkiest subgroup; it has been conceived as a single central body on which the mobile planes (flap, ailerons) and the front leading edge are assembled. The mobile planes are assembled aside in workplace 9, while the front leading edge is assembled on the bench. For the wing slip dimensions we have established a length of 12.2 m and a height of 1.5 m, Figure 5.51.

Figure 5.49. Workplace of right and left frustum tail assembly.

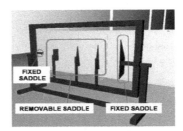

Figure 5.50. Back frustum in detail.

Figure 5.51. Workplace 5: assembly of the wing.

5.11.5 Workplace 6: fins

In workplace 6 the fins are assembled, Figure 5.52. The workplace consists of a workbench and an assembly slip of the subgroup, composed of a base on which the upper body is supported and welded. The body is formed of a structure on which six couples of mobile saddles are lodged, which serve as references for the positioning of the interior ribs of the drift and which can be open or closed in relation to the phase of assembly, Figure 5.53.

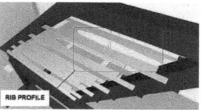

Figure 5.52. Workplace of fin assembly.   **Figure 5.53. Assembly slip of the fin.**

The assembly slip has a height of about 2.4 m and a length of 3.3 m. For this reason, a staircase has been positioned in the workplace in order to allow the operators to work in the upper part of the slip. The main activities carried out in this workplace are: drilling and riveting. As in the previous workplace, the assembly kit is withdrawn from the store, the presence of all the components is checked and subsequently the assembly activities begin. Once the assembly is completed, the quality control of the subgroup and the subsequent testing are evaluated .

5.11.6 Workplace 7: rudder

Workplace 7 is dedicated to the assembly of the rudder. Here the assembly slip of the subgroup and a work bench are arranged. The dimensions of the tool are smaller than those used for the drifts; in fact it has a maximum height of 2.2 m and length of 2.1 m, Figure 5.54-5.55. The assembly sequence of the rudder, like the other subgroups, has not yet been defined, but we have proposed a sequence in order to optimize the assembly cycle and to define the operations that the workers have to perform, satisfying ergonomic requirements, as described in the previous section.

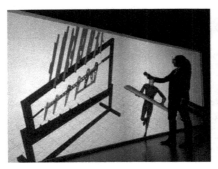 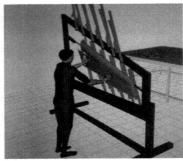

Figure 5.54. Interior structure of the rudder .   **Figure 5.55. Workplace of the rudder .**

5.11.7 Workplace 8: horizontal stabilizer

In workplace 8 the horizontal stabilizer is assembled. Here the assembly slip and a workbench are arranged. The assembly slip has a length of around 5 m and a height of 2.20 m. It consists of a rigid body and a base on which the body is welded. A series of transversal elements called saddles are connected to the body; they represent the references for positioning the ribs and the longerons, Figure 5.56. The saddles perfectly represent the profile of the ribs; they are mobile and rotate around the fixing pivot of the slip, in order to allow the operators all the necessary operations for the assembly, Figure 5.57. The assembly activities of this subgroup, like the others, begin when the assembly kits are brought to the workplace, where the presence of all the necessary components for the assembly is checked.

5.11.8 Workplace 9: flaps and ailerons

The flaps and the ailerons are assembled in a single workplace where two assembly slips are placed. The slips have been designed with the same logic as the tail plane of the aeroplane. Both consist of two main structures: the base and the body that rotates around its principal axis. Saddles are positioned on the body that serve as references for the positioning of the ribs.

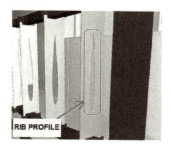

Figure 5.56. Workplace of stabilizer assembly.   **Figure 5.57. Assembly slip stabilizer.**

5.11.9 Workplace 10: final assembly

Workplace 10 is dedicated to the final assembly of the fuselage. In fact, once the assembly of the main subgroups is completed, the wing, the front and the central frustum are slung and moved to workplace 10 with a bridge crane. The others parts, since they are less bulky and lighter, are transported manually to the same workplace. In workplace 10 the support for the final assembly is placed, consisting of a rigid structure with five saddles that reproduce the profile of the fuselage, and a couple of saddles in the final part that reproduce the profile of the two frustums of the tail, Figure 5.58. Note that the final part of the structure support is split; this allows assembly operations between the frustums of the tail. In fact, the operators can easily access the space between the two frustums of the tail in order to perform all the operations necessary. First the front frustum and then the central frustum are positioned and fixed to the slip, and the two frustums are joined. In order to carry out the joining operations between the bodies, stairs have been positioned also in the upper part of the fuselage, that can be moved manually by the operators and positioned in accordance with the joining zone of the two frustums. This allows the operator to perform all the necessary operations in an easier way, Figure 5.59. Once the two parts are joined, the two frustums of the tail are positioned and joined to the rest of the fuselage. At this point, the carts and the drifts are assembled on the fuselage. The wing, the horizontal plan of the tail, the flaps and the ailerons, that are assembled on the whole fuselage and not on the single wing, have to be assembled. In this workplace the assembly cycle of the structure of the aircraft is concluded, Figure 5.60.

Subsequently, the structure is removed from the support and placed in the following workplace where the assembly of all the equipment and systems is carried out.

 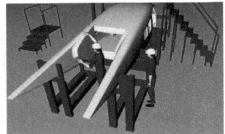

Figure 5.58. Fuselage frustum assembly.   **Figure 5.59. Tail frustum positioning.**

5.12   CONCLUSIONS AND FUTURE WORK

To date, light aircraft process development methodologies have remained traditional, if compared with those of major aeronautical firms. In recent years, the main objective of light aircraft industries has been (and continues to be) to reconcile high quality and a high level of innovation with low costs of development and production. This work represents a first step in this direction. Through virtual reality software and CAD tools it is possible to merge the development and the engineering processes of the aircraft. Usually one of the most difficult problems to be resolved is to reconcile the product project with the most proper methodologies for its realization.

This work describes a possible approach for the development of new methodologies for planning the production, that aims at reconciling the designer with the production manager and therefore at matching the product design to the manufacture. In this way, by simulating both mechanical and human performance in a virtual environment, it is possible to anticipate, and consequently to avoid, critical errors in the assembly process design. In future, the same methodology could be applied to a more detailed and in depth study of the subgroups of the aircraft, like the wing and the frustums, that are the most critical due to the massive structure of the elements of which they are composed and due to their complexity. This methodology could also be applied to the assembly of the fittings and to the definition of the paths of all the cables on the aircraft. A new methodology could be studied in detail in order to appraise the possibility of assembling some or all of the cables of each subgroup in the assembly workplaces.

Figure 5.60. "Sky Car"aircraft.

5.13 REFERENCES

[5.1] Various Authors: Research and developed department firema trasporti. descrizione dell'attività produttiva. Internal documentation, pp 1–6 (2003)

[5.2] Di Gironimo, G.: Study and development of ergonomic design methodologies in a virtual environment. PhD thesis, University of Naples Federico II, Naples (2002)

[5.3] Various Authors: Handbook Jack 4(1), (2005)

[5.4] Various Authors: On line Guide Jack 4(1), (2005)

[5.5] Abshire, K.J., Barron, M.K.: Virtual mainteinance: Real world application within virtual environments. In: Proceedings of Realiability and Maintainability Symposium, Ohio (1998)

[5.6] Caputo, F., Di Gironimo, G., Patalano, S., Biondi, A., Liotti, F., Maione, R.: Use of ergonomic software in virtual environment for the prevention of muscular–skeletal diseases. In: Proceedings of ISCS 2001, Naples (2001)

[5.7] Di Gironimo, G., Patalano, S., Liotti, F.: A new approach to the evaluation of the risk due to manual material handling through digital human modelling. In: Proceedings of ISCS 2003, Cefalù (2003)

[5.8] Caputo, F., Di Gironimo, G., Monacelli, G., Sessa, F.: The design of a virtual environment for ergonomics studies. In: Proceedings of XII ADM International Conference, Rimini, Italy (2001)

[5.9] Di Gironimo, G., Monacelli, G., Martorelli, M., Vaudo G.: Use of virtual mock-up for ergonomic design. In: Proceedings of 7th International Conference on "The role of experimentation in the automotive product development process". ATA, Florence (2001)

[5.10] Caputo, F., Di Gironimo, G., Fiore, G., Patalano, S.: Maintainability and ergonomic tests of locomotive design through virtual mainikins (in italian). In: Proceeding of Convegno Nazionale XIV ADM e XXXIII AIAS "Innovazione nella Progettazione Industriale". (2004)

[5.11] Gerbino, S.: Interoperability among cad systems. In: Proceedings of XIIIADM–XVIngegraf InternationalConference. Napoli, Italy (2003)

[5.12] Various Authors: Website ABB. http://www.abb.it

[5.13] Di Gironimo G., Marzano A, 2006. Design of an Innovative Assembly Process of a Modular Train in Virtual Environment. In Proceeding of Virtual Concept 2006, Playa del Carmen, Mexico, (selected paper for the International Journal on Interactive Design and Manufacturing, Vol. 1 No. 2 June 2007 pp.85-97).

[5.14] VV.AA, 2003. Research & Developed Department OMA SUD Sky Technology. Descrizione dell'attività produttiva. Internal documentation, 2003.

[5.15] Abshire K.J., Barron M.K., 1998. Virtual Mainteinance: Real World Application within Virtual Environments. Proc. of Realiability and Maintainability Symposium, Ohio, 1998.

[5.16] Caputo F., Di Gironimo G., Patalano S., Biondi A., Maione R., Liotti F., 2001. Use of Ergonomic Software in Virtual Environment for the Prevention of Muscular-skeletal Diseases. Proc. of ISCS 2001, Naples.

[5.17] Di Gironimo G., 2002. Study and development of ergonomic design methodologies in virtual environment (in Italian: Studio e sviluppo di metodologie di progettazione ergonomica in ambiente virtuale). Ph.D Dissertation, University of Naples Federico II, Naples.

[5.18] Di Gironimo G., Patalano S., Liotti F., 2003. A New Approach to the Evaluation of the Risk due to Manual Material Handling through Digital Human Modelling. Proc. of ISCS 2003, Cefalù.

[5.19] Caputo F., Di Gironimo G., Fiore G., Patalano S., 2004. Verifiche di manutenzione e di ergonomia nella progettazione di una locomotiva di manovra mediante l'impiego di modelli virtuali. Atti del Convegno Nazionale XIV ADM e XXXIII AIAS "Innovazione nella Progettazione Industriale", Bari.

[5.20] Caputo F., Di Gironimo G., Marzano A., 2006. Ergonomic optimization of a manufacturing system work cell in a virtual environment, Acta Polytechnica Journal, vol.46 No. 5/2006. pp.21-27.

[5.21] Di Gironimo G., Monacelli G., Martorelli M., Vaudo G., 2001. Use of Virtual Mock-Up for Ergonomic Design. Proc. of 7th International Conference on "The role of experimentation in the automotive product development process" – ATA 2001. Florence, Italy.

[5.22] Caputo F., Di Gironimo G., Monacelli G., Sessa F., 2001. The Design of a Virtual Environment for Ergonomics Studies. Proceedings of XII ADM International Conference, Rimini, Italy.

[5.23] Di Gironimo G., Monacelli G., Patalano S., 2004. A Design Methodology for Maintainability of Automotive Components in Virtual Environment. In Proceeding of International Design Conference – Design 2004, Dubrovnik.

CHAPTER 6

HUMAN ROBOT INTERACTION IN VIRTUAL REALITY

6.1 INTRODUCTION

In this chapter, we describe a framework which helps designers visualize and verify the results of robotic work cell simulation in a Virtual Environment (VE). The system aims at significantly reducing production costs and error sources during manufacturing processes. The means to achieve these goals are the development of a prototypical VE for the support of robots planning tasks, reuse of animation events, and the implementation of customization tools for animation elements and their behaviour. By using advanced Virtual Reality (VR) techniques, the system is also able to direct the focus of the observer to interesting events, objects and time-frames during robotic simulations in order to highlight the Human Robot Interaction within the manufacturing systems.

In recent years industrial engineering has been oriented towards the development of flexible manufacturing systems, [6.1] and in particular man-machine interaction systems. Research in robotics is looking for different applications where a human being is to be conceived not exclusively as an operator programming off-line the robot, but rather as a system interacting with the machine by means of different modes, [6.2]. The Virtual Reality (VR) technology offers a highly potential in terms of planning and development of manufacturing systems, [6.3], [6.4], [6.5]. In this work, we propose a new research methodology that uses VR techniques in the field of the so-called Anthropic Robotics. Anthropic Robotics refers to the study of the technologies and methodologies to develop automatic machines that operate services in environments cohabited with the humans, such as cooperating robots, [6.6]. The robotic systems, that work in the same environment of humans, have to be endowed with strong characteristics of autonomy, reliability and safety. Indeed, they have to be able to respond to breakdowns, collisions or any unexpected change of the operational scenario and to be able to guarantee the human safety at the same time.

The robotic work cells are important elements in automated manufacturing systems for delivering required manufacturing materials and operations with industrial robots and associated peripheral devices. Rapid design and deployment of a robotic work cell require the successful applications of concepts, tools, and methods for fast product design, manufacturing process planning, and plant floor cell control support. An important technology for achieving this goal is robotic work cell simulation. Robotic work cell simulation is a modelling-based problem solving approach that aims to sufficiently produce credible solutions for robotic system design. Figure 6.1 shows a simple robotic work cell simulation model. The current practices of robotic work cell design have proven that successfully implementing robotic simulation brings the designers many benefits, [6.7], [6.8], [6.9], [6.10].

First, designing a robotic work cell via robotic simulation eliminates the guesswork from concepts and unrealistic expectations based on technical equipment specifications. Designers can offer optimum solutions via having evaluated alternatives. Second, as modifications are made to a work cell design, the process of incorporating modifications into the corresponding work cell simulation models is easier and faster, compared to making changes to a real work cell. Finally, robotic simulation packages bring designers a safe design environment. Whether, designing a new work cell, optimizing its performance or making modifications to an operational work cell and developing and testing required programs that can be safely carried out.

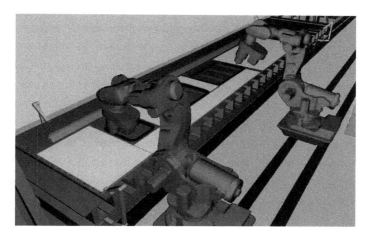

Figure 6.1. Robotic work cell.

Models in robotic work cell simulation are principles for studying the behaviour of the actual work cell devices over time. These models may be geometric objects, mathematical equations and relations, or graphical representations. Designers usually use the commercial robotic simulation software packages to build simulation models. The simulation design tasks require designers to constitute specific methods and procedures via selecting and executing the functions provided by the simulation packages with appropriate design data. This process often involves both inductive and deductive reasoning and requires multifaceted knowledge in diverse disciplines such as computer-aided design (CAD), machine design, and robotics, [6.11]. The robotic simulation designers often face significant theoretical and technical challenges in

123

understanding and applying the current robotic simulation technology. To deal with this challenge, this paper introduces a methodology for developing the required robotic work cell simulation models via virtual reality (VR) technology.

The significance of visualisations in manufacturing simulations has been discusses by [6.10]. Visualisation of simulations in 3D supports validation and verification of the model. It helps analysts to understand the simulation results. Since manufacturing simulations can be inherently complex, results of manufacturing simulations can also be complex and counter-intuitive. By watching the areas of interest, the designer can see what is happening and understand how the dynamic behaviour of the system affects the results. Through visualisations, results can also be better communicated to a non-technical audience and give it more credibility. However, while the advantages of simulation visualisations are clear, a number of issues have to be overcome to make it a common method to be deployed in discrete simulation projects. Since simulation models and graphical representations of systems in virtual environments are entirely different, the graphical model has to be developed in addition to the simulation model. Without the right tools and methods this will not only add a significant amount of effort to the overall costs of a simulation project, but can also lead to inconsistencies between the simulation model and the virtual scene and in turn introduce new error sources into the simulation process. To make the creation process of simulation visualisations effective, reuse of object models and animation behaviour is mandatory. It reduces costs for creation of models and implementation of animation methods. It also improves the credibility of the animated virtual model because its components have been deployed and tested before. Inconsistencies between the virtual model and the simulation model, especially during the phase of refinement, can be avoided by binding simulation parameters and properties of animation elements together. Once a parameter and a property are bound, they have the same value and any changes on one of them will be propagated to the other. This technique even supports the configuration of the simulation model up to a certain extent through changes in the virtual scene.

As mentioned before, results of manufacturing simulations are prone to be complex. One advantage of presenting these results in a virtual scene is that information can be better presented in their context. However, virtual representations of systems can become large and advanced visualisation techniques are required to give the user orientation in time and space.

6.2   RELATED WORK

In [6.12] a system framework founded on animation elements is described. The animation elements are used to visualize discrete events, e.g. discrete simulation results [6.13]. [6.14] discusses methods for rapid generation of animation elements. Animation Elements encompass geometric description of its visual appearance and adaptable animation behaviour. Animation Elements have an object-oriented design and emphasis was put on reusability and on overloading of the behaviour methods. [6.15] introduces the concept of 3D components and 3D frameworks. Instead of conventional software objects, software components were now used to encapsulate visual appearance and animation behaviour. The advantage here is that the animation components could now be visually composed into larger scenes.

In [6.16] the concept of component-based visualisation of simulation results was introduced. A tool was described, which could analyse simulation results and visualize them. No interaction with the simulation system was possible and the application could not visualize real-time data.

6.3 DYNAMIC SHARED OBJECT FOR ROBOT CONTROL

The VD2 computational engine does not provide native support for defining and handling kinematic chains. For this reason we developed a DSO module, called *robot.so*, to manage open kinematic chain manipulators in Virtual Reality. The main goals are:

- The plug-in has to be flexible, so that the same functions have to be suitable for different types of robot;
- The robot has to be able to reproduce an user-defined path;
- The user has to be able to manage the robot in real-time;
- Any eventual end-of-stroke condition has to be signalled to the system.

6.3.1 The robot hierarchical model

In order to use the functions provided by the DSO module, the first step is to arrange the geometric model of the robot. In general, a kinematic chain is a set of rigid elements, called links, connected by joints. A joint is essentially a constraint on the geometric relationship between two adjacent links. Since VD2 does not provide a really constraint-based Virtual Environment, the scene-graph tree structure has been used to keep the logical sequence of the different links. Thus, each joint of the chain is represented by an assembly node, as shown in Figure 6.2

Thanks to the hierarchical structure of the scene graph, the programmer does not have to be concerned about the numerical solution of the direct kinematic problem. Indeed, a single geometric transformation, such as a rotation about an axis, can be defined for a specific joint of the chain, without concerns about the configuration of the other joints. However, since the tree structure of the scene-graph is acyclic, the hierarchical model described above is only suitable for open kinematic chains.

.

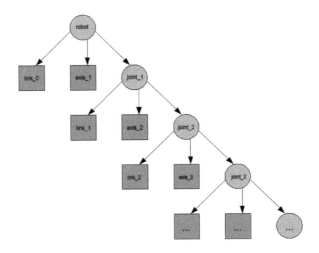

Figure 6.2. The robot hierarchical model.

6.3.2 Robot Configuration file

One of the most important goals is the flexibility of the module: in other words the DSO module should be able to handle different types of kinematic chain, independently from number and type of the axes the robot is equipped with. In order to achieve this, the kinematic chain has to be described in a configuration file, which specifies not only names of joints and axes, as defined in the robot scene-graph, but also type (revolute or prismatic) and working range of each axis, see Figure 6.3

```
  IRB7600.cfg  ×
# IRB7600 configuration file

# Sintax:
# ROBOT       = int      Robot ID
# JOINT       = string   Name of the joint assembly node, as defined in the scene-graph
# AXIS        = string   Name of the axis geometry, as defined in the scene-graph
# QMAX, QMAX  = float    Axis working range
# TYPE        = int      Joint type (1=REVOLUTE, 2=PRISMATIC)
# Q0          = float    Initial joint position.

ROBOT = 0
{

   JOINT joint_1
   {
       TYPE = 1;
       QMIN = -180;
       QMAX = 180;
       Q0 = 0;
       AXIS = axis_0;
   }
   JOINT joint_2
   {
       TYPE = 1;
       QMIN = -60;
       QMAX = 80;
       Q0 = 0;
       AXIS = axis_1;
   }
   JOINT joint_3
   {
       TYPE = 1;
```

Figure 6.3. An industrial robot configuration file.

126

6.3.3 Robot task planning

Early industrial robots were programmed by moving the robot to a desired goal point and recording its positions in a memory, which the sequencer would read during playback. During teaching phase, the user can guide the robot directly by hand, or through the interaction with a teach pendant, that is a hand-held control terminal which allows the user to move each joint of the manipulator, [6.17]. The DSO module provides functions to simulate in Virtual Reality both aforementioned teaching systems. Indeed, it is possible to control the kinematic chain through the flystick, that is a wireless interaction device designed especially for VR applications (see chapter 2), or to make the robot follow a tracked object, such as the virtual hand.

Moreover, set of actions can be specified in the scene description file, in order to carry out the robotic simulation as realistically as possible.

6.3.4 Path planning

The DSO module allows to define different postures for the robot, by specifying each of them in the scene description file. Moreover, it is possible to handle the kinematic chain in real-time, by relating an input from a VR device, such as a button of the flystick, to the handling of a specific joint, see Figure 6.4.

Figure 6.4. Real-time path planning.

In this way, the user causes the robot to assume the desired posture, by using the flystick as a teach-pendant. Each posture can be stored in a file, so that the user can define a point-to-point path. The reproduction of the defined path then can be triggered by an event, as it happens for any other action in the VE. Thus, the features described above provide an easy way to plan a collision-free path for the robot task. Furthermore, the integration with the underlying software architecture also allows the user to plan quite complex behaviours, so that the robot can manipulate objects or manage any eventual collision, see Figure 6.5.

6.3.5 Direct end-effector positioning

Generally, finding the joint angles for a given position of the end-effector in the operational space requires an inverse kinematic approach, as described in [6.18]. Since this analysis is limited to open kinematic chain manipulators, a simpler but effective methodology has been adopted. As aforementioned, VD2 provides a specific action that cause a virtual object to follow the user hand, within a specified

constraint. For instance, an object can only rotate about a defined axis, according to the movement of the virtual hand. Unfortunately, each constraint is related to a single object in the scene-graph and it is treated separately from the other constraints. In other words, the user cannot define directly kinematic relationships among two or more virtual objects. This approach is suitable for modeling simple kinematics, such as a virtual door, but it can lead to an unexpected behaviour when it is applied to a kinematic chain, because generally each link of the chain will move independently from the other elements, as illustrated in Figure 6.6.

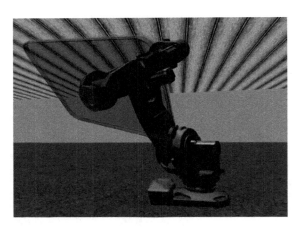

Figure 6.5. The manipulator reproducing an assembly task.

In order to avoid the breaking of the kinematic chain, each constraint operates on a different joint-node of the hierarchical model described in section 6.4.1, rather than directly on the geometries of each link. For instance, according to the kinematic chain shown in Figure 6.7, the first joint-node contains the whole kinematic chain, the second includes only the last two links and finally the third node is just the last link of the chain.

128

Figure 6.6. The manipulator reproducing an assembly task.

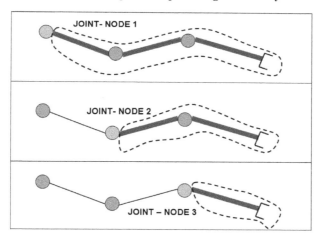

Figure 6.7. The kinematic chain breaks during the constrained movement.

Since all the geometries belonging to a specific joint-node act as a single "rigid object" during the movement, the geometric relationships among the different links will be kept in any case.

Many constraints can be triggered by a single event, such as the collision between the virtual hand and a specific link of the robot. In this way, the user can drag the whole kinematic chain by "grasping" the end-effector until the robot reaches the desired position, see Figure 6.8. At the same time, it is also possible to move by hand only one or more links of the chain.

129

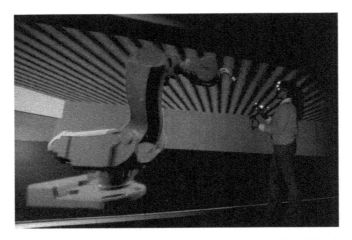

Figure 6.8. The kinematic chain being dragged by the virtual hand.

6.3.6 Final considerations

As aforementioned, the library functions allow the user to easily plan an intended task for any type of open kinematic chain manipulator: it is only necessary to prepare the robot scene-graph and then edit the configuration file according to the type of chain. Since the DSO module is completely integrated with the underlying software framework, the user can take benefit from all the others functionality provided by the

Simulation Manager. For instance, it is possible to display the working area of the robot, highlight eventual collisions between the robot and any object in the VE or trace the path of the end-effector (sweeping) during the task execution, see Figure 6.9. The basic command set can also be used to simulate specific robot behaviours triggered by certain events, eventually generated by external modules. Thus, the modular approach adopted by VD2 kernel could be used to interface a real sensor network with a virtual robot cell. In this way, it would be possible to test the safety strategies adopted to control robots that operate in anthropic domains.

However, the plug-in has some limitations:

- The user cannot define a path in the operational space directly, because the library is not able to perform inverse kinematic operations. Indeed, the solution of the inverse kinematic problem would break the requirements of flexibility of the DSO module, since it is strictly depending on the particular robot type. Moreover, it would need to consider also eventual kinematic redundancy issues, [6.19];

- The module can manage only open kinematic chain manipulators;

- A dynamic model of the kinematic chain has not been implemented.

Figure 6.9. The sweeping action applied to the end-effector.

Furthermore, in order to use the function set provided by the DSO module, the user has to prepare the robot scenegraph and the related configuration file manually. Thus, a future goal will be to develop a graphic wizard to lead the user through the configuration process.

6.4 A VIRTUAL-REALITY-BASED EVALUATION ENVIRONMENT FOR WHEELCHAIR-MOUNTED MANIPULATORS

The design of solutions for robotic extenders of wheelchairs must take into account both objective and subjective metrics for everyday activities in human environments.

Virtual Reality (VR) constitutes a useful tool to effectively test design ideas and to verify performance criteria. This paper presents the development of a simulation environment, where three different manipulators to be mounted on a commercially available wheelchair have been considered. Experimental results are discussed in a significant case study, based upon users' feedback.

Research in the field of assistive applications is playing a key role in the international robotics community. Several research groups are developing systems aimed at assisting disabled people in the actions and assignments typical of everyday life in both structured and unstructured domestic environments [6.20]. The main goal of robotics for assistance is to increase the quality of life of disabled people; in particular, robot manipulators are required, able to replicate human abilities in terms of strength, speed and accuracy in the manipulation of objects and tools. While such systems can offer autonomy to impaired persons, great challenges are presented by the study of the interface, the suitability of available robotic systems for special users, the usability, especially related to the kind of disability. While wheelchair-mounted manipulators are

131

becoming common [6.21], [6.22], realistic simulation tools for studying their use are required. The design of human-robot collaboration tools has to pay particular attention to the following issues:

- user's safety,
- system ergonomics and usability,
- cost-effectiveness.

The study of the aforementioned issues requires design tools able to simulate not only the robotic system and its control interface, but also unexpected behaviours in anthropic environments, depending both on the user and the system, such as the occurrence of mechanical/electronics failures or unexpected user movements within the robot workspace. Moreover, a realistic interface can be helpful for appreciating the cognitive Human–Robot Interaction (cHRI). The main advantage of the immersive Virtual Reality (VR) technology in the field of human–robot interaction

is the ability to simulate the aforementioned dynamic events [6.23], [6.24]. Hence, the objective of this paper is to demonstrate the effectiveness of a VR-based simulator (Figure 6.10) for testing the usability and possible applications of wheelchair–mounted robot manipulators, with an effective solution for mounting available robot manipulators on commercial wheelchairs via a sliding rail (Figure 6.11). The objective validation of the safety measures necessary to provide a dependable human-robot cooperation is becoming central for service robotics [6.25], [6.26], [6.27]: these evaluations can be complemented with measures of systems usability that are provided via realistic simulation experiments.

Finally, manipulators which almost replicate the kinematic structure of the human arm can be chosen for the legibility of their motion, which could improve the confidence of the users during physical Human-Robot Interaction (pHRI).

Figure 6.10. Wheelchair-mounted manipulator simulation in virtual environment.

132

6.4.1 Wheelchair-mounted manipulators

Assistive robots can be divided in fixed structures and moving platforms. While the first solution often requires modifications of the infrastructures in order to provide a known environment, manipulators mounted on mobile vehicles or on wheelchairs offer higher flexibility. A good discussion of these issues has been addressed in [6.20]. It is worth noticing that often disabled people with upper-limb limitations also present mobility impairments which force them to use wheelchairs. A wheelchair-mounted manipulator [6.21], [6.22], [6.28] can be an effective extender but, on the other hand, realistic simulations of the environment and extensive experimental activities have to be conducted for testing the effectiveness of their applications in unstructured domains. Moreover, safety issues have to be addressed in depth [6.25]. The use of Virtual Reality could speed up the design of such robotics solutions, because it is possible to set systems parameters based on feedback from experimenters, involving also cognitive aspects of the interaction with the robots. Such instrument can be used for a fast comparison of interface, appearance, kinematic parameters. The research work described in this paper has consisted of two stages; namely, a concept stage in which the functional parameters of the wheelchair and the robotic arm have been defined, and an evaluation stage in which a VR architecture has been developed in order to analyze the safety issues and to evaluate the usability of the system.

6.4.2 Concept Stage

The integrated system has to guarantee the maximum effectiveness and usability. At the same time, the introduction of the robotic arm does not have to require any significant change on its surrounding environment. Moreover, the wheelchair-mounted manipulator has to satisfy the following requirements:

- reduced weight,
- intrinsic safety towards accidental collisions with the user.

The powered wheelchair Indoor 2003 [6.29] has been chosen, based on consideration on its features, which are reported in Figure 6.12.

In order to obtain a wider workspace for a robotic extender mounted on the wheelchair, a sliding rail has been considered around the powered wheelchair, with proper modelling of such a joint for exploiting it as an additional degree of freedom (DOF) available for robot control. The manipulator can move around the wheelchair by sliding along the rail; the rail is able to rotate around an horizontal axis, providing a way to change its inclination, for adapting the workspace to user's needs (e.g., better dexterity on the ground). Such characteristic widely increases the robot workspace.

Figure 6.11. Digital wheelchair integrated with manipulator and sliding rail.

Three different lightweight robot arms (Figure 6.13), have then been considered for integration with the wheelchair:

a. KUKA Light Weight Robot (LWR),

b. Amtec Ultra Light Weight Robot (ULWR),

c. Mitsubishi PA-10.

These manipulators have a kinematic structure similar to the human arm: moreover, the reduced weight allows using them for service robotics, while quantitative evaluation of intrinsic safety in case of rigid impacts are available only for the KUKA arm [6.26].

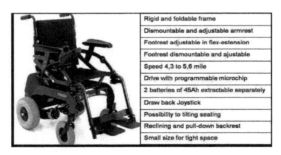

Figure 6.12. The physical indoor wheelchair.

134

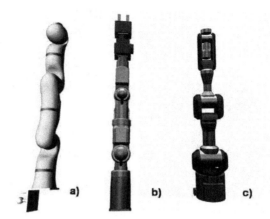

Figure 6.13. The compared manipulators are: (a) KUKA LWR, (b) Amtec ULWR, (c) Mitsubishi PA-10.

Since the robotic arm and the rail mounted on the wheelchair introduce static balancing issues, it has been necessary to verify also the stability of the integrated system, which has been verified for all the three robots considering the wheelchair both with and without a person sitting on it.

6.4.3 VR architecture for evaluation

The experimental activity has been carried out using VR technologies in the laboratory of the Competence Centre for the Qualification of Transportation Systems, set-up by Campania Regional Authority in Caserta, (Figure 6.14). The Simulation Manager software is Virtual Design 2, by vrcom. This application provides a user-friendly interface to handle all VR devices used in the laboratory, as described in chapter 3. Moreover, the availability of a Software Development Kit allows enhancing and customizing the basic functionalities of the Simulation Manager with new software modules, that are fully integrated with the underlying VR framework. In order to simulate the movement of a kinematic chain in VR, a new module has been developed (see section 6.3).

Such plug-in allows the user to handle a robotic arm in real-time through a 3D input device, such as the joystick or the space-mouse, or the flystick.

Figure 6.14. The immersion is a key issue for capturing subjective evaluation about the virtual model.

Since the library is also capable of inverse kinematics operations, the kinematic chain can be handled both in the joint and the operational space. The user can provide reference positions or velocities to the robot. Each posture can be saved and reproduced in real-time. Moreover, the possibility to display the manipulability ellipsoid and the Jacobian matrix related to a certain posture is given (Figure 6.16)

Therefore, the control module can be adapted to every kind of manipulators, where proper static analysis and evaluation about the risks of collisions with such manipulators allow using it for assistance tasks.

In order to enhance the illusion of using a real appendix of a wheelchair, a physical wheelchair has been placed in the laboratory in a way such that the user viewpoint coincided with the virtual wheelchair starting position. The user can move the wheelchair in the virtual space, by means of the flystick. Moreover, in order to increase the immersion feeling, the shutter glasses are endowed with optical targets, and the user can also adjust the point of view on the virtual scene by moving the head. The robotic arm can be moved by means of a joystick. The joystick is a 3D input device, but it has only 4 DOFs. Therefore, the joystick only handles the positional component of the end-effector, and consequently any orientation adjustments have to be done by operating in the joint space. For this reason, the user can individually control each joint angle by means of the joystick buttons. In alternative, it is possible to use a 6-DOF input device, such as the space-mouse, but this choice has been discarded mainly due to the difficulty for training a disabled user in controlling both the position and the orientation of the end-effector at the same time.

136

Moreover, each button of the joystick is related to a predefined posture. This feature simplifies the handling of the kinematic chain, since generally it reduces the time needed to reach a desired posture. Finally, a 2D-Menu allows the operator to select the robot to be evaluated and a set of test environments, as described in the following. It is worth noticing that other kinds of input devices can be tested with minor modifications to the interface. The second step of the experimental phase is the training of the users. The experimenter has to be briefly trained to the use of the interaction devices and also informed about the importance of their experience [6.30]. The training stage has been carried out in two phases: in a first phase, the operator has simply to move the end-effector along the three axes of the virtual space; then, in a second phase, the operator has to move the end-effector following a defined path. Only once the user is able to use properly the virtual devices, the experimental session can begin. In order to compare the proposed robots for the considered application, an appropriate methodology of analysis has been developed.

Five parameters have been chosen for comparison: manipulability, usability, safety (simplified, based on robot's weight), payload, and design. The manipulability denotes the attitude of the robot to perform a defined assignment starting from the current configuration. It is an intrinsic positional characteristic of the manipulator [13]. The usability denotes the ease the user can handle the kinematic chain. It strictly depends on the robot type, the inverse kinematics algorithm, the user-robot interface, and the level of training of the user. In order to quantify the usability for a specific robot, we defined a set of time-based experiments. For the considered tasks, the usability measure is based on the elapsed time for completing the required motion. The safety measure considers the weight of the robot and the power needed for its operation, based on nominal data provided by the manufacturer. A more comprehensive safety criterion of a robotic system should include several factors such as: robot weight, impact measure, dependability of control algorithm, operator skill independence, environmental factors.

The payload of the robot can become important because it allows disabled users to transport objects in the house: it is significant to relate it to the robot's weight The indicator about the appreciation of the design is a number in the range [1,5] which takes into account color, shape, smoothness of the manipulator design. The importance of the robot design on the emotional reactions of the user is also important [6.31]. Such parameter is obviously subjective and it is influenced by the age and cultural level of the user.

The second step has been the set-up of some time-based tests aimed at measuring these parameters. with the aim to underline the potentialities and the weaknesses of the three manipulators and to provide a judgment on the efficiency of the robot control algorithm.

Four tasks have been conceived:

1. book positioning on a shelf (reaching);

2. objects relocation on a shelf (dexterity);

3. moving a chess piece on a board (fine motion between obstacles);

4. moving objects between two different planes (free motion);

1) Book positioning on a shelf: Twelve books of four different colours are positioned on four different shelves. The task consists of moving every book on the shelves of the same colour. This test allows determining the feasibility of operations performed by disabled people, as well to measure the manipulability and the usability of the robot for different levels of height of the end-effector.

Figure 6.15. Object relocation on a shelf.

2) Objects relocation on a shelf: The user is 0.4 m far from a shelf of height 1 m, where five equidistant markers are set. The distance among the most external markers is set to 1 m. It deals with the normal dimension of the human arm workspace. A spherical object is set on the central marker. The user first has to grab the object from the central marker and then release it on the others. The test evaluates

the horizontal usability of the robot (Figure 6.15).

3) Moving a chess piece on a board: The user is 0.380m far from the chessboard (measured by the chest to the centre of the board). The violet horse (Figure 6.16) is the piece to be moved, the yellow box represents the starting position of the horse and blue boxes highlight the final destination of the piece. The user has to move the piece from the yellow box to the blue one. This test evaluates the ability to handle small objects among some obstacles.

4) Moving objects between two different planes: In this test, the user has to move some glasses from a table to another and back again. The two tables are 0,9 m far. The starting position and the final destination of the glasses are signalled by different coloured markers.

6.4.4 Experimental results

The objective measures and the more subjective indicators suggested for evaluating the wheelchair-mounted manipulators have to be combined in a measure of user friendliness of the considered systems.

Figure 6.16. Moving a chess piece: the manipulability ellipsoid is displayed on the screen.

Experimental sessions have been performed by able bodied users, while weights for suggesting the judgement criteria have been collected among potential disabled users

(whose opinions have been collected via the Internet [6.32]), and robotics experts. The following results are related to the chessboard task. A measure of the presented indicators has been computed for each manipulator. The numeric values are arranged in the 3×5 matrix E, whose generic $e_{i,j}$ element is a number indicating the performance for the i-th robot related to the j-th criterion.

Then, the following normalization method has been adopted for all the indicators: each element $e_{i,j}$ is divided to the sum of the numbers corresponding to the i-th performance of all the three robots, and the result is indicated as $e_{0i;j}$.

The 3×5 matrix E_0 whose generic row i is the vector with the 5 normalized performance scores $e_{0i;j}$, $j=1:5$, is called the impact matrix, reported in Table I.

139

Then, the results have been weighted based on multicriteria decision analysis techniques [6.33]. In this second phase, E_0 is multiplied for a weighting score matrix W, which indicates users' feedback about the importance that they give to the proposed criteria. In particular, the following weighting factors in the matrix W have been collected via interviews (Table II), where a generic column labelled with the name of an user contains the level of importance assigned by that user to the five proposed indicators. For the experiments, four impaired users (indicated with $U_{j,j} = 1$; 4), and two robotics experts (E_1, E_2) have been considered.

Table 6.1. Impact values for the proposed performance indicators.

	Manip.	Usability	Design	Payload	Safety
KUKA	0.21	0.34	0.44	0.54	0.47
Amtec	0.31	0.33	0.34	0.1	0.33
Mitsubishi	0.48	0.33	0.22	0.36	0.2

Finally, the performance of the manipulators is summarized in the appraisal score matrix S, namely,

$$S = E\,W \quad (1)$$

which is reported in Table III. It is clear that the analysis can be done also with a reduced number of indicators, in order to appreciate the effect of some single performance metrics on the overall score.

Table 6.2. Weighting values (ranging from 1 to 10) giving importance to the proposed criteria according to possible users (u) and experts (e).

	U1	U2	U3	U4	E1	E2
Manipulability	6	7	7	8	6	8
Usability	9	10	10	7	9	7
Design	7.5	8	8	6	8	7
Payload	10	6	7	9	6	7
Safety	7	4	8	10	9	10

These preliminary results suggest the potential impact of the proposed tool for the evaluation of robotic extenders: with increasing users and criteria, the design can focus on some solutions which are preferred by the users. In addition, the indicator of safety has to be refined: the total weight can be misleading, since the inertia at the collision instants can be different, e.g., based on the configuration assumed by different robots for the same task. Moreover, the judgment on the design is practically independent on the task. Different indicators can be considered as well depending on the application: the possibility of carrying heavy objects, e.g., is considered important by the participants, for autonomy in their houses, while the central requirement of safety is considered somehow less central, due to the will of taking the control of the robot without autonomous robot behaviours. It is worth noticing that some weights chosen by the users can be quite surprising for an engineer. This shows that the appearance of the robot has a strong impact on the user, and even the intrinsic safety and versatility can be appreciated not enough, if not accompanied by a proper design.

Table 6.3. Appraisal scores for the considered robots.

	U1	U2	U3	U4	E1	E2
KUKA	19.37	16.91	19.33	18.64	18.37	18
Amtec	13.66	13.41	14.83	13.34	14.09	13.48
Mitsubishi	15.47	14.68	15.84	15.02	14.54	14.52

6.6 CONCLUSION AND FUTURE WORK

The proposed simulation environments can allow comparison between trajectory planning schemes, kinematic optimization of robots for wheelchairs, appearance and

reliability of wheelchair-mounted manipulators. With the addition of simplified dynamic models, joint torques due to the interaction with the environment can be generated as well for a better modelling of the environment. The use of robots in unstructured and time-varying environments implies the need for implementing real-time reactive strategies to cope with possible collisions [6.34], which are going to be implemented for completing the simulator. The proposed tool can be used also for evaluating in a very realistic way the reactions during the approach and the motion of the robot on desired or unexpected trajectories. Force feedback can be added via a proper haptic interface [6.35]. Finally, tests with disabled persons could provide additional insights in the cognitive and ethical aspects related to the introduction of robotic extenders in the everyday life. The study on the interface should take into account the possible difficulty for a disabled user in controlling both the position and the orientation of the end-effector at the same time.

Virtual reality provides a time- and cost-effective tool for the proposed comparisons.

6.7 REFERENCES

[6.1] Craig J.: Simulation-based robot cell design in adeptrapid. In Proceeding of the 1997 IEEE International Conference on Robotics and Automation, ICRA, Albuquerue (Apr. 1997), vol. 4, pp. 3214–3219.

[6.2] Chen D., Cheng F.: Integration of product and process development using rapid prototyping and work cell simulation technologies. Journal of Industrial Technology 16, 1 (2000), 2–5.

[6.3] Caputo F., Di Gironimo G., Marzano A.: Approach to simulate manufacturing systems in virtual environment. In Proc. of the XVIII Congreso International de Ingeniería Gráfica (May 2006).

[6.4] Caputo F., Di Gironimo G., Marzano A.: Ergonomic optimization of a manufacturing system work cell in a virtual environment. In Proc. of 5th International Conference on Advanced Engineering Design (June 2006). Selected paper for the Acta Polytechnica Journal, vol. 46 No. 5/2006 pp. 21-27.

[6.5] De Amicis R., Di Gironimo G., Marzano A.: Design of a virtual reality architecture for robotic work cells simulation. In Proceeding of Virtual Concept 2006, Playa del Carmen, Mexico (Nov. 2006).

[6.6] Alami R., Albu-Schaeffer A., Bicchi A., Bischoff R., Chatila R., De Luca A., De Santis A., Giralt G., Guiochet J., Hirzinger G., Ingrand F., Lippiello V., Mattone R., Powell D., Sen S., Siciliano B., Tonietti G., Villani L.: Safe and dependable physical human-robot interaction in anthropic domains: State of the art and challenges. In IROS 2006 IEEE/RSJ International Conference on Intelligent Robots and Systems. Workshop on Physical Human-Robot Interaction in Anthropic Domains, Beijing (Chine) (Oct. 2006). http://www.laas.fr/~felix/publis/pdf/iros06ws.pdf.

[6.7] Rudnick F.C., Moore S.G. Robotic paint simulation and off-line programming in a rapid prototyping environment. In Proceedings of Robotic simulation & off-line programming workshop, Seattle, Washington, 1997.

[6.8] Knasinski A.B. Linking Simulation to the real world trough robot metrology. In Proceeding of Robotic Simulation & Off-line Programming Workshop, Seattle, Washington, 1997.

[6.9] Craig J.J. Simulation-based robot cell design in AdeptRapid. In Proceeding of the 1997 IEEE International Conference on Robotics and Automation, ICRA, Albuquerue, April, Vol.4. pp. 3214-3219, 1997.

[6.10] Rohrer M. W. Seeing is believing: the importance of visualization in manufacturing simulation. In Proceeding of the 2002 winter simulation conference.

[6.11] Chen D., Cheng F. Integration of product and process development using rapid prototyping and work cell simulation technologies. Journal of Industrial Technology, Vol.16, No. 1. pp. 2-5, 2000.

[6.12] Luckas V. Elementbasierte, effiziente und schnelle generierung von 3D visualisierungen und 3D animationen. Dissertation at the school of computer engineering, Darmstadt University of Technology, Darmstadt, 2000.

[6.13] Luckas V., Broll T. CASUS – An object- oriented three-dimensional animation system for event-oriented simulators. In Proceeding of Computer Animation' 97, pp. 144-150, Geneva.

[6.14] Dorner R., Elcacho C., Luckas V. Behaviour authoring for VRML application in industry. Technical Notes of Eurographics'98, Lisbon, 1998.

[6.15] Dorner R., Grimm P. Building 3D applications with 3D components and 3D frameworks. Workshop structured design of virtual environments. At WEB3D, Paderborn, 2001.

[6.16] Quick J. M. Component based 3D visualization of simulation results. Advanced simulation technologies conference, San Diego, California, 2002.

[6.17] Craig J. J.: Introduction to Robotics: Mechanics and Control. Prentice Hall, 2003.

[6.18] Zhao J., Badler N. I.: Inverse kinematics positioning using nonlinear programming for highly articulated figures. ACM Transactions on Graphics 13, 4 (1994), 313–336.

[6.19] Sciavicco L., Siciliano B.: Robotica industriale- Modellistica e controllo di manipolatori. McGraw-Hill, 2000.

[6.20] A. Gimenez, C. Balaguer, A. Sabatini, V. Genovese, "The MATSsystem to assist disabled people in their home environments", 2003 IEEE/RSJ International Conference on Intelligent Robots and Systems, Las Vegas, NV, USA, 2003.

[6.21] H. Eftring, K. Boschian, "Technical results from manus user trials", 1999 International Conference on Rehabilitation Robotics, Stanford, CA, USA, 1999.

[6.22] M. Hillman, A. Gammie, "The Bath institute of medical engineering assistive robot", 1994 International Conference on Rehabilitation Robotics, Wilmington, DE, USA, 1994.

[6.23] G. Burdea, P. Coiffet, Virtual Reality Technology, (2nd Ed.), Wiley– Interscience, 2003.

[6.24] G. Burdea, P. Coiffet, "Virtual reality and robotics," in Handbook of Industrial Robotics, 2nd ed., S. Nof, Ed. New York: Wiley, 1999, ch. 17, pp. 325-333.

[6.25] R. Alami, A. Albu-Sch¨affer, A. Bicchi, R. Bischoff, R. Chatila, A. De Luca, A. De Santis, G. Giralt, G. Hirzinger, V. Lippiello, Mattone, S. Sen, B. Siciliano, G. Tonietti, L. Villani

"Safe and dependable physical human-robot interaction in anthropic domains: state of the art and challenges", Workshop on Physical Human-Robot Interaction, 2006 IEEE/RSJ Conference on Intelligent Robots and Systems, Beijing, PRC, 2006.

[6.26] S. Haddadin, A. Albu-Sch"affer, G. Hirzinger, "Dummy crashtests for evaluation for rigid human-robot impacts", 2007 IARP International Workshop on Technical Challenges for Dependable Robots in Human environments, Rome, I, 2007.

[6.27] Y. Ota, T. Yamamoto, "Standardization activity for service robotics", 2007 IARP International Workshop on Technical Challenges for Dependable Robots in Human environments, Rome, I, 2007.

[6.28] R.M. Alqasemi, E.J. McCaffrey, K.D. Edwards, R.V. Dubey, "Wheelchair-mounted robotic arms: Analysis, evaluation and development", 2005 IEEE/ASME International Conference on Advanced Intelligent Mechatronics, Monterey, CA, USA, 2005.

[6.29] www.neatech.it.

[6.30] G. Di Gironimo, M. Guida, "Developing a Virtual Training system in Aeronautical Industry", 5th EUROGRAPHICS Italian Chapter Conference, Trento, I, 2007.

[6.31] http://robocare.istc.cnr.it

[6.32] www.disabiliforum.com

[6.33] Voogd, H., Multicriteria Evaluation for Urban and Regional Planning, Pion, London, UK, 1983.

[6.34] A. De Santis, A. Albu-Sch"affer, C. Ott, B. Siciliano, and G. Hirzinger, "The skeleton algorithm for self-collision avoidance of a humanoid manipulators", 2007 IEEE-ASME International Conference on Advanced Intelligent Mechatronics, Zurich, CH, 2007.

[6.35] Burdea, G., Force and Touch Feedback for Virtual Reality, Wiley, New York, USA, 1996.

APPENDIX

Optimum solution report

WELDING VISUAL CONTROL

Static Strength Prediction

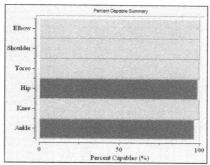

5th PERCENTILE

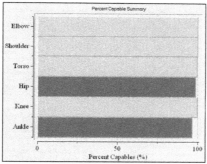

50th PERCENTILE

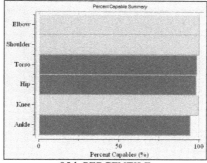

95th PERCENTILE

147

OWAS

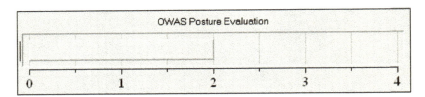

5th PERCENTILE

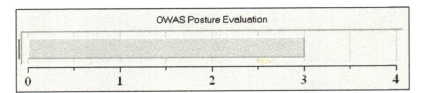

50th PERCENTILE

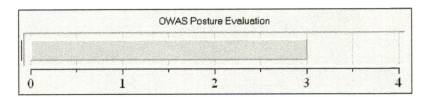

95th PERCENTILE

Low Back Analysis

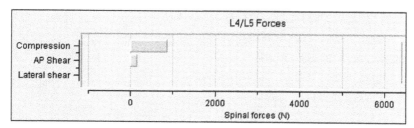

5th PERCENTILE

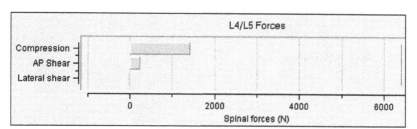

50th PERCENTILE

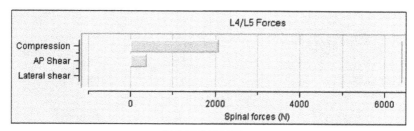

95th PERCENTILE

149

Body Group A Posture Rating

Upper arm: 2
Lower arm: 3
Wrist: 3
Wrist Twist: 2
Total: 4

Muscle Use: Normal, no extreme use
Force/Load: < 2 kg intermittent load
Arms: Supported

Body Group B Posture Rating

Neck: 1
Trunk: 3
Total: 3

Muscle Use: Normal, no extreme use
Force/Load: < 2 kg intermittent load

Legs and Feet Rating

Standing, weight even. Room for weight changes.

Grand Score:

Body Group A Posture Rating

Upper arm: 2
Lower arm: 2
Wrist: 3
Wrist Twist: 2
Total: 4

Muscle Use: Normal, no extreme use
Force/Load: < 2 kg intermittent load
Arms: Supported

Body Group B Posture Rating

Neck: 1
Trunk: 3
Total: 3

Muscle Use: Normal, no extreme use
Force/Load: < 2 kg intermittent load

Legs and Feet Rating

Standing, weight even. Room for weight changes.

Grand Score:

Body Group A Posture Rating

Upper arm: 3
Lower arm: 3
Wrist: 3
Wrist Twist: 2
Total: 5

Muscle Use: Normal, no extreme use
Force/Load: < 2 kg intermittent load
Arms: Supported

Body Group B Posture Rating

Neck: 1
Trunk: 3
Total: 3

Muscle Use: Normal, no extreme use
Force/Load: < 2 kg intermittent load

Legs and Feet Rating

Standing, weight even. Room for weight changes.

Grand Score:

WELDING IMPERFECTION RESTORING WITH BRUSH

Static Strength Prediction

153

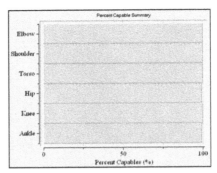

5th PERCENTILE

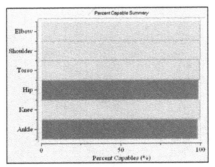

50th PERCENTILE

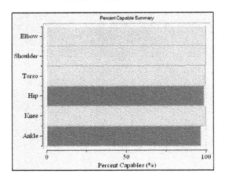

95th PERCENTILE

154

OWAS

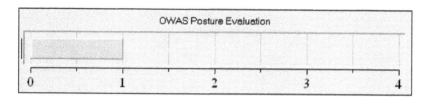

5th PERCENTILE

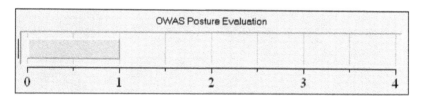

50th PERCENTILE

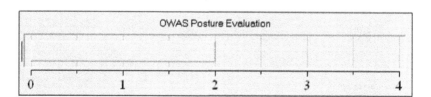

95th PERCENTILE

Low Back Analysis

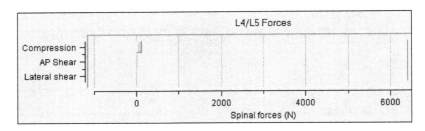

5th PERCENTILE

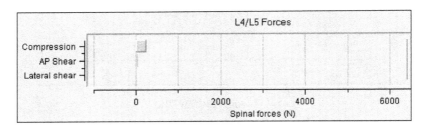

50th PERCENTILE

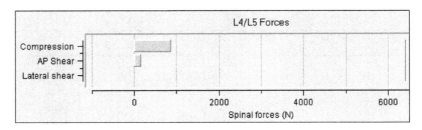

95th PERCENTILE

RULA

Body Group A Posture Rating

Upper arm: 4
Lower arm: 3
Wrist: 2
Wrist Twist: 2
Total: 5

Muscle Use: Normal, no extreme use
Force/Load: < 2 kg intermittent load
Arms: Not supported

Body Group B Posture Rating

Neck: 1
Trunk: 3
Total: 3

Muscle Use: Normal, no extreme use
Force/Load: < 2 kg intermittent load

Legs and Feet Rating

Standing, weight even. Room for weight changes.

Grand Score:

50th PERCENTILE

157

Body Group A Posture Rating

Upper arm: 4
Lower arm: 3
Wrist: 2
Wrist Twist: 2
Total: 5

Muscle Use: Normal, no extreme use
Force/Load: < 2 kg intermittent load
Arms: Not supported

Body Group B Posture Rating

Neck: 1
Trunk: 3
Total: 3

Muscle Use: Normal, no extreme use
Force/Load: < 2 kg intermittent load

Legs and Feet Rating

Standing, weight even. Room for weight changes.

Grand Score:

95th PERCENTILE

158

Body Group A Posture Rating

Upper arm: 3
Lower arm: 3
Wrist: 3
Wrist Twist: 2
Total: 5

Muscle Use: Normal, no extreme use
Force/Load: < 2 kg intermittent load
Arms: Supported

Body Group B Posture Rating

Neck: 1
Trunk: 3
Total: 3

Muscle Use: Normal, no extreme use
Force/Load: < 2 kg intermittent load

Legs and Feet Rating

Standing, weight even. Room for weight changes.

Grand Score:

BRAZE WELDING RENEWAL

Static Strength Prediction

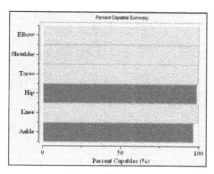

5th PERCENTILE

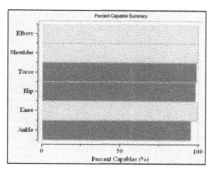

50th PERCENTILE

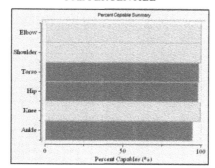

95th PERCENTILE

OWAS

161

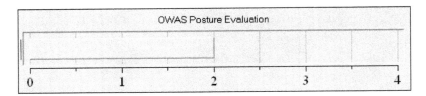

5th PERCENTILE

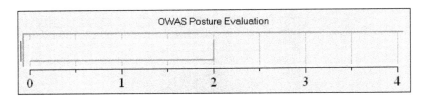

50th PERCENTILE

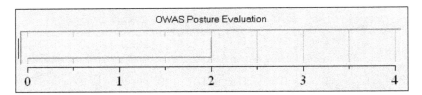

95th PERCENTILE

Low Back Analysis

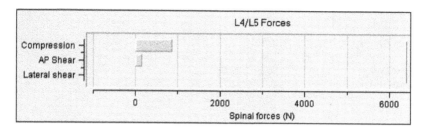

5th PERCENTILE

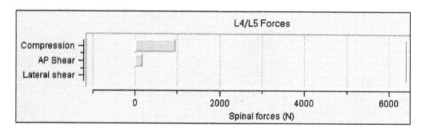

50th PERCENTILE

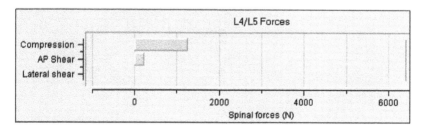

95th PERCENTILE

RULA

163

Body Group A Posture Rating

Upper arm: 4
Lower arm: 2
Wrist: 3
Wrist Twist: 2
Total: 5

Muscle Use: Normal, no extreme use
Force/Load: < 2 kg intermittent load
Arms: Not supported

Body Group B Posture Rating

Neck: 1
Trunk: 2
Total: 2

Muscle Use: Normal, no extreme use
Force/Load: < 2 kg intermittent load

Legs and Feet Rating

Standing, weight even. Room for weight changes.

Grand Score:

Body Group A Posture Rating

Upper arm: 3
Lower arm: 3
Wrist: 3
Wrist Twist: 2
Total: 5

Muscle Use: Normal, no extreme use
Force/Load: < 2 kg intermittent load
Arms: Not supported

Body Group B Posture Rating

Neck: 1
Trunk: 2
Total: 2

Muscle Use: Normal, no extreme use
Force/Load: < 2 kg intermittent load

Legs and Feet Rating

Standing, weight even. Room for weight changes.

Grand Score:

95th PERCENTILE

Body Group A Posture Rating

Upper arm: 2
Lower arm: 2
Wrist: 3
Wrist Twist: 2
Total: 4

Muscle Use: Normal, no extreme use
Force/Load: < 2 kg intermittent load
Arms: Not supported

Body Group B Posture Rating

Neck: 1
Trunk: 2
Total: 2

Muscle Use: Normal, no extreme use
Force/Load: < 2 kg intermittent load

Legs and Feet Rating

Standing, weight even. Room for weight changes.

Grand Score:

SMEARING SEALER WITH GUN

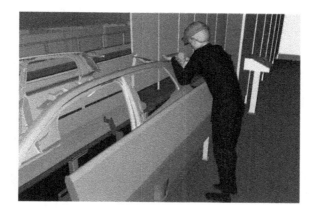

Static Strength Prediction

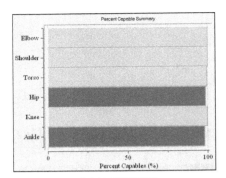

5th PERCENTILE

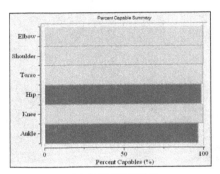

50th PERCENTILE

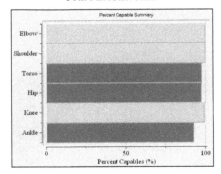

95th PERCENTILE

OWAS

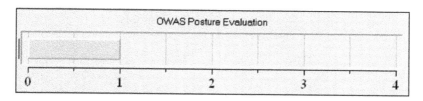

5th PERCENTILE

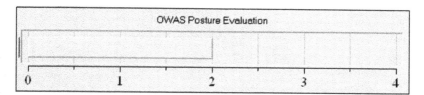

50th PERCENTILE

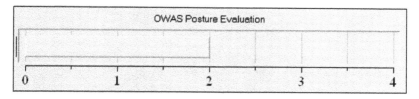

95th PERCENTILE

Low Back Analysis

169

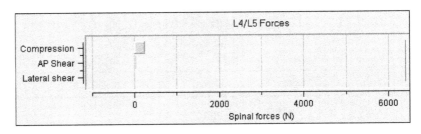

5th PERCENTILE

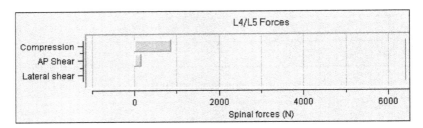

50th PERCENTILE

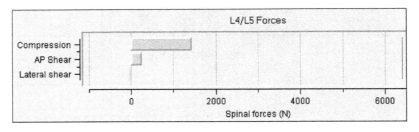

95th PERCENTILE

RULA

Body Group A Posture Rating

Upper arm: 4
Lower arm: 2
Wrist: 3
Wrist Twist: 2
Total: 5

Muscle Use: Normal, no extreme use
Force/Load: < 2 kg intermittent load
Arms: Not supported

Body Group B Posture Rating

Neck: 1
Trunk: 2
Total: 2

Muscle Use: Normal, no extreme use
Force/Load: < 2 kg intermittent load

Legs and Feet Rating

Standing, weight even. Room for weight changes.

Grand Score:

Body Group A Posture Rating

Upper arm: 3
Lower arm: 3
Wrist: 3
Wrist Twist: 2
Total: 5

Muscle Use: Normal, no extreme use
Force/Load: < 2 kg intermittent load
Arms: Not supported

Body Group B Posture Rating

Neck: 1
Trunk: 2
Total: 2

Muscle Use: Normal, no extreme use
Force/Load: < 2 kg intermittent load

Legs and Feet Rating

Standing, weight even. Room for weight changes.

Grand Score:

95th PERCENTILE

Body Group A Posture Rating

Upper arm: 3
Lower arm: 3
Wrist: 3
Wrist Twist: 2
Total: 5

Muscle Use: Normal, no extreme use
Force/Load: < 2 kg intermittent load
Arms: Not supported

Body Group B Posture Rating

Neck: 1
Trunk: 3
Total: 4

Muscle Use: Mainly static, e.g. held for longer than 1 minute
Force/Load: < 2 kg intermittent load

Legs and Feet Rating

Standing, weight even. Room for weight changes.

Grand Score: 5

SMEARING SEALER RENEWAL

Static Strength Prediction

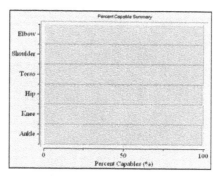

5th PERCENTILE

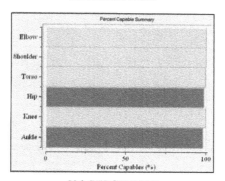

50th PERCENTILE

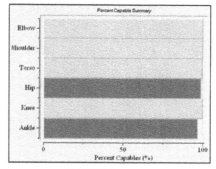

95th PERCENTILE

OWAS

175

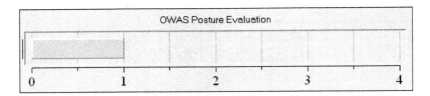

5th PERCENTILE

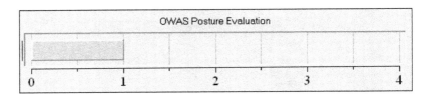

50th PERCENTILE

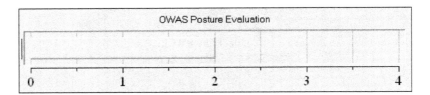

95th PERCENTILE

Low Back Analysis

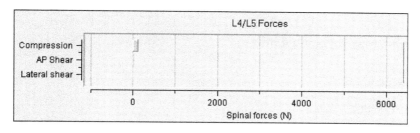

5th PERCENTILE

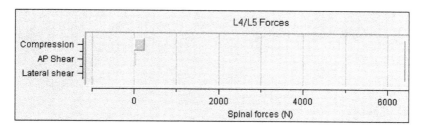

50th PERCENTILE

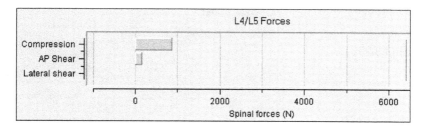

95th PERCENTILE

RULA

Body Group A Posture Rating

Upper arm: 4
Lower arm: 2
Wrist: 3
Wrist Twist: 2
Total: 5

Muscle Use: Normal, no extreme use
Force/Load: < 2 kg intermittent load
Arms: Not supported

Body Group B Posture Rating

Neck: 1
Trunk: 2
Total: 2

Muscle Use: Normal, no extreme use
Force/Load: < 2 kg intermittent load

Legs and Feet Rating

Standing, weight even. Room for weight changes.

Grand Score:

Body Group A Posture Rating

Upper arm: 3
Lower arm: 3
Wrist: 3
Wrist Twist: 2
Total: 5

Muscle Use: Normal, no extreme use
Force/Load: < 2 kg intermittent load
Arms: Not supported

Body Group B Posture Rating

Neck: 1
Trunk: 2
Total: 2

Muscle Use: Normal, no extreme use
Force/Load: < 2 kg intermittent load

Legs and Feet Rating

Standing, weight even. Room for weight changes.

Grand Score:

95th PERCENTILE

Body Group A Posture Rating

Upper arm: 4
Lower arm: 3
Wrist: 3
Wrist Twist: 2
Total: 6

Muscle Use: Mainly static, e.g. held for longer than 1 minute
Force/Load: < 2 kg intermittent load
Arms: Supported

Body Group B Posture Rating

Neck: 1
Trunk: 1
Total: 2

Muscle Use: Mainly static, e.g. held for longer than 1 minute
Force/Load: < 2 kg intermittent load

Legs and Feet Rating

Standing, weight even. Room for weight changes.

Grand Score:

BOTTOM CROSS MEMBER WINDSCREEN LOADING

Static Strength Prediction

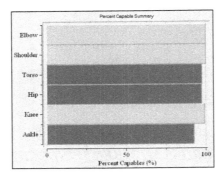

5th PERCENTILE

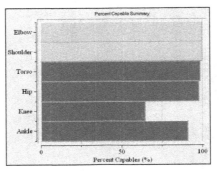

50th PERCENTILE

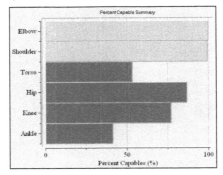

95th PERCENTILE

OWAS

182

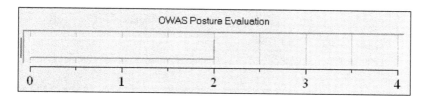

5th PERCENTILE

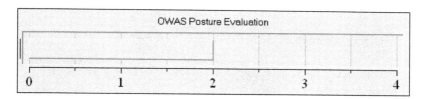

50th PERCENTILE

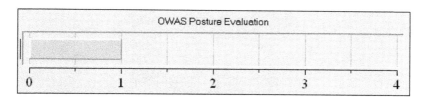

95th PERCENTILE

Low Back Analysis

183

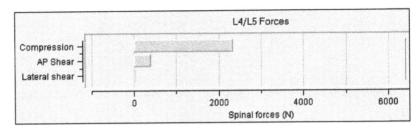

5th PERCENTILE

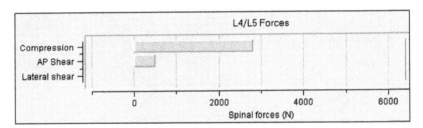

50th PERCENTILE

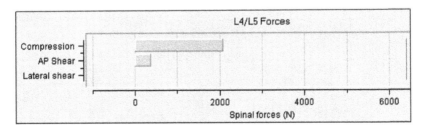

95th PERCENTILE

RULA

Body Group A Posture Rating

Upper arm: 5
Lower arm: 3
Wrist: 3
Wrist Twist: 2
Total: 9

Muscle Use: Normal, no extreme use
Force/Load: 2-10 kg static load or 2-10 kg repeated load
Arms: Not supported

Body Group B Posture Rating

Neck: 6
Trunk: 5
Total: 11

Muscle Use: Normal, no extreme use
Force/Load: 2-10 kg static load or 2-10 kg repeated load

Legs and Feet Rating

Standing, weight even. Room for weight changes.

Grand Score: 7

Body Group A Posture Rating

Upper arm: 5
Lower arm: 3
Wrist: 3
Wrist Twist: 2
Total: 9

Muscle Use: Normal, no extreme use
Force/Load: 2-10 kg static load or 2-10 kg repeated load
Arms: Not supported

Body Group B Posture Rating

Neck: 6
Trunk: 5
Total: 11

Muscle Use: Normal, no extreme use
Force/Load: 2-10 kg static load or 2-10 kg repeated load

Legs and Feet Rating

Standing, weight even. Room for weight changes.

Grand Score: 7

Body Group A Posture Rating

Upper arm: 5
Lower arm: 3
Wrist: 3
Wrist Twist: 2
Total: 9

Muscle Use: Normal, no extreme use
Force/Load: 2-10 kg static load or 2-10 kg repeated load
Arms: Not supported

Body Group B Posture Rating

Neck: 6
Trunk: 5
Total: 11

Muscle Use: Normal, no extreme use
Force/Load: 2-10 kg static load or 2-10 kg repeated load

Legs and Feet Rating

Standing, weight even. Room for weight changes.

Grand Score: 7

UPPER CROSS MEMBER WINDSCREEN LOADING

Static Strength Prediction

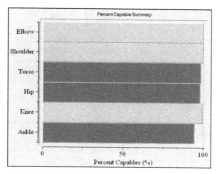

5th PERCENTILE

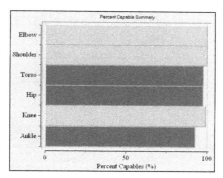

50th PERCENTILE

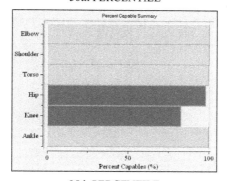

95th PERCENTILE

OWAS

189

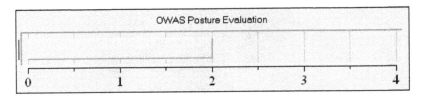

5th PERCENTILE

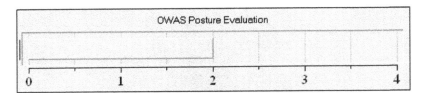

50th PERCENTILE

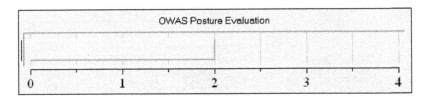

95th PERCENTILE

Low Back Analysis

190

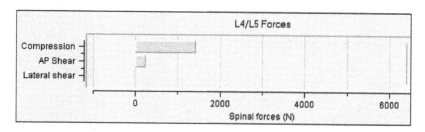

5th PERCENTILE

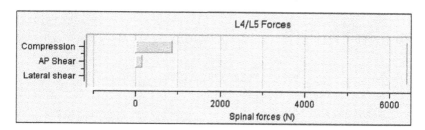

50th PERCENTILE

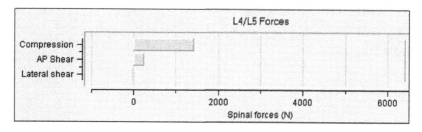

95th PERCENTILE

RULA

191

Body Group A Posture Rating

Upper arm: 4
Lower arm: 3
Wrist: 2
Wrist Twist: 2
Total: 6

Muscle Use: Normal, no extreme use
Force/Load: 2-10 kg intermittent load
Arms: Not supported

Body Group B Posture Rating

Neck: 1
Trunk: 2
Total: 3

Muscle Use: Normal, no extreme use
Force/Load: 2-10 kg intermittent load

Legs and Feet Rating

Standing, weight even. Room for weight changes.

Grand Score: 5

Body Group A Posture Rating

Upper arm: 4
Lower arm: 3
Wrist: 2
Wrist Twist: 2
Total: 6

Muscle Use: Normal, no extreme use
Force/Load: 2-10 kg intermittent load
Arms: Not supported

Body Group B Posture Rating

Neck: 1
Trunk: 2
Total: 3

Muscle Use: Normal, no extreme use
Force/Load: 2-10 kg intermittent load

Legs and Feet Rating

Standing, weight even. Room for weight changes.

Grand Score: 5

95th PERCENTILE

Body Group A Posture Rating

Upper arm: 4
Lower arm: 3
Wrist: 2
Wrist Twist: 2
Total: 6

Muscle Use: Normal, no extreme use
Force/Load: 2-10 kg intermittent load
Arms: Not supported

Body Group B Posture Rating

Neck: 1
Trunk: 2
Total: 3

Muscle Use: Normal, no extreme use
Force/Load: 2-10 kg intermittent load

Legs and Feet Rating

Standing, weight even. Room for weight changes.

Grand Score: 5

BACK CROSS MEMBER WINDSCREEN LOADING

Static Strength Prediction

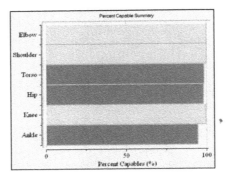

5th PERCENTILE

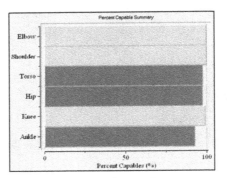

50th PERCENTILE

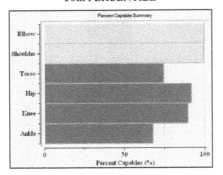

95th PERCENTILE

OWAS

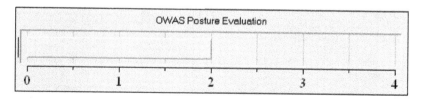

5th PERCENTILE

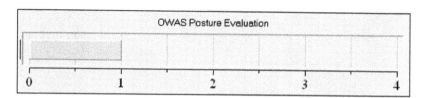

50th PERCENTILE

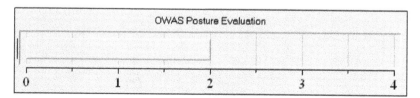

95th PERCENTILE

Low Back Analysis

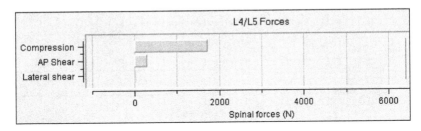

5th PERCENTILE

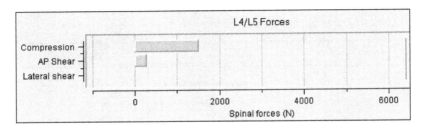

50th PERCENTILE

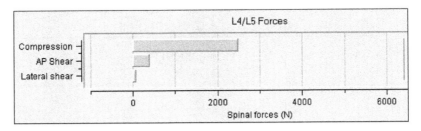

95th PERCENTILE

RULA

Body Group A Posture Rating

Upper arm: 4
Lower arm: 3
Wrist: 3
Wrist Twist: 1
Total: 6

Muscle Use: Normal, no extreme use
Force/Load: 2-10 kg intermittent load
Arms: Not supported

Body Group B Posture Rating

Neck: 1
Trunk: 3
Total: 4

Muscle Use: Normal, no extreme use
Force/Load: 2-10 kg intermittent load

Legs and Feet Rating

Standing, weight even. Room for weight changes.

Grand Score: 6

Body Group A Posture Rating

Upper arm: 4
Lower arm: 3
Wrist: 2
Wrist Twist: 2
Total: 5

Muscle Use: Normal, no extreme use
Force/Load: < 2 kg intermittent load
Arms: Not supported

Body Group B Posture Rating

Neck: 1
Trunk: 3
Total: 3

Muscle Use: Normal, no extreme use
Force/Load: < 2 kg intermittent load

Legs and Feet Rating

Seated, Legs and feet well supported. Weight even.

Grand Score:

95th PERCENTILE

Body Group A Posture Rating

Upper arm: 4
Lower arm: 3
Wrist: 2
Wrist Twist: 2
Total: 6

Muscle Use: Normal, no extreme use
Force/Load: 2-10 kg intermittent load
Arms: Not supported

Body Group B Posture Rating

Neck: 1
Trunk: 3
Total: 4

Muscle Use: Normal, no extreme use
Force/Load: 2-10 kg intermittent load

Legs and Feet Rating

Standing, weight even. Room for weight changes.

Grand Score: 6

UPPER FRONT RAFTER LOADING

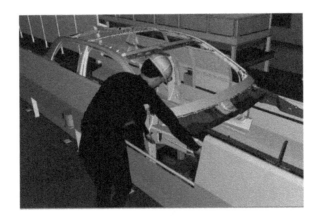

Static Strength Prediction

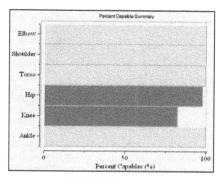

5th PERCENTILE

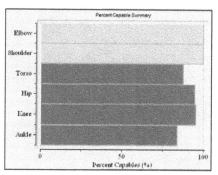

50th PERCENTILE

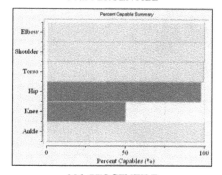

95th PERCENTILE

OWAS

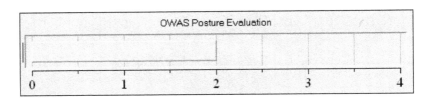

5th PERCENTILE

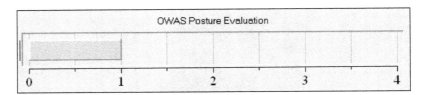

50th PERCENTILE

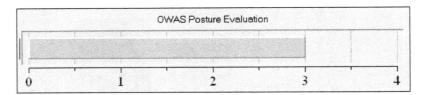

95th PERCENTILE

Low Back Analysis

204

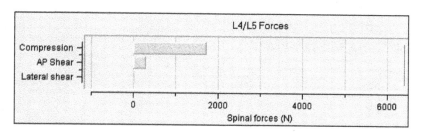

5th PERCENTILE

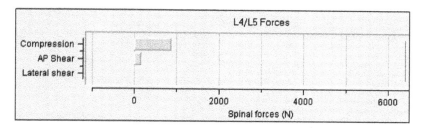

50th PERCENTILE

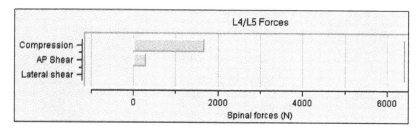

95th PERCENTILE

RULA

205

Body Group A Posture Rating

Upper arm: 3
Lower arm: 3
Wrist: 3
Wrist Twist: 1
Total: 5

Muscle Use: Normal, no extreme use
Force/Load: 2-10 kg intermittent load
Arms: Not supported

Body Group B Posture Rating

Neck: 1
Trunk: 3
Total: 4

Muscle Use: Normal, no extreme use
Force/Load: 2-10 kg intermittent load

Legs and Feet Rating

Standing, weight even. Room for weight changes.

Grand Score: 5

Body Group A Posture Rating

Upper arm: 4
Lower arm: 3
Wrist: 2
Wrist Twist: 2
Total: 5

Muscle Use: Normal, no extreme use
Force/Load: < 2 kg intermittent load
Arms: Not supported

Body Group B Posture Rating

Neck: 1
Trunk: 3
Total: 3

Muscle Use: Normal, no extreme use
Force/Load: < 2 kg intermittent load

Legs and Feet Rating

Seated, Legs and feet well supported. Weight even.

Grand Score:

95th PERCENTILE

Body Group A Posture Rating

Upper arm: 4
Lower arm: 3
Wrist: 2
Wrist Twist: 2
Total: 5

Muscle Use: Normal, no extreme use
Force/Load: < 2 kg intermittent load
Arms: Not supported

Body Group B Posture Rating

Neck: 1
Trunk: 3
Total: 3

Muscle Use: Normal, no extreme use
Force/Load: < 2 kg intermittent load

Legs and Feet Rating

Seated, Legs and feet well supported. Weight even.

Grand Score: